Guides: 38

TIDEPOOL AND NEARSHORE FISHES
OF CALIFORNIA

by

JOHN E. FITCH

and

J. LAVENBERG

by Evie Templeton

1406

CALIFORNIA PRESS
ANGELES · LONDON

COUNTY LIBRARY
CALIFORNIA

California Natural History Guides
Arthur C. Smith, General Editor

Advisory Editorial Committee:
A. Starker Leopold
Robert Ornduff
Robert C. Stebbins

Dedicated to
EARL STANNARD HERALD
April 10, 1914, to January 16, 1973
and
ROLF LING BOLIN
March 22, 1901, to August 23, 1973

University of California Press
Berkeley and Los Angeles, California
University of California Press, Ltd.
London, England

Copyright ©1975, by
The Regents of the University of California

ISBN 0-520-02844-9 (clothbound)
ISBN 0-520-02845-7 (paperbound)
Library of Congress Catalog Card Number: 74-84145

Printed in the United States of America

Acknowledgments

In writing a book such as this, one realizes how much he depends upon his fellow man for assistance. Specimens must be collected and studied, taxonomic problems worked out, and drawings and photographs made; and once a manuscript has been drafted it still must be typed, reviewed, and edited.

During the past 25 years innumerable sportfishermen, commercial fishermen, skindivers, and biologists have contributed fishes from which we have obtained basic information pertinent to a factual account such as this. Among those who have provided specimens, life-history data, distributional information, catch statistics, and taxonomic expertise are: E. A. Best, John Bleck, Richard C. Burge, W. L. Craig, Lillian Dempster, W. I. Follett, Herbert Frey, Jerry Goldsmith, Daniel W. Gotshall, Carl L. Hubbs, Robert N. Lea, Kim McCleneghen, Daniel J. Miller, Basil Nafpaktitis, Lawrence Quirollo, Jack W. Schott, Steven A. Schultz, Camm Swift, Boyd W. Walker, and Parke H. Young.

Some of those who accompanied us on various fossil-collecting trips and helped with the digging, rock-splitting, and screening and sorting of residues are: Lloyd W. Barker, Richard A. Fitch, Richard Huddleston, and Mark Roeder.

Reeve Bailey, John Bleck, William Eschmeyer, Jerry Neumann, and Richard Rosenblatt permitted us to examine institutional collections entrusted to their care.

Mary Butler, Larry Reynolds, Armando Solis, and Evie Templeton prepared the drawings, and Bruce Collette, C. Richard Robins, Richard Rosenblatt, and Boyd Walker, experts in fish identification, reviewed them.

Cover and color-plate photographs were taken by: Jack Ames, James Coyer, John Duffy, Jack Engle, Robert Given, Daniel Gotshall, Ed Hobson, Edwin Janss, Jr., Preston Smith, John Stephens, Charles Turner, and Edward Zimbelman.

Helen Abraham, Dorothy Fink, Terri Kato, and Sherri Owings typed various sections and drafts of the manuscript, and Arline Fitch and Jerry Neumann each read it several times and made helpful suggestions.

We are especially grateful to the California Department of Fish and Game and Natural History Museum of Los Angeles County for permitting access to their reference libraries, personnel, fish collections, and photographic files. If we have overlooked any individual or institution, it has not been intentional.

CONTENTS

INTRODUCTION

The shallow waters immediately adjacent to California's 1200 miles of coastline offer a wide variety of habitats where at least 250 kinds of fishes live out all or some of their lives. Most are small, brightly-colored fishes which seldom travel any great distance, either on a given day or throughout a lifetime. The eggs and larvae of some may drift seaward and lead a pelagic existence, but once their wanderlust is satisfied, the juveniles return to nearshore areas where they take up life as their parents did before them. A few kinds wander into shallow water in search of food, to spawn, or perhaps out of curiosity, but not to take up residence.

Although food, temperature, salinity, oxygen, vegetation, substratum, shelter, companionship, currents, and tides are probably important to all fishes, each kind has its own recipe for mixing these ingredients in order to come up with an ideal situation. To the blind goby, for instance, oxygen, salinity, and tides are secondary to its need to find the burrow of a particular kind of shrimp, the only habitat in which it can survive. The garibaldi cannot spawn unless it can locate and culture specific kinds of red algae in which to hatch its eggs. Grunion depend upon sandy beaches and certain kinds of high tides for successful egg-laying and subsequent hatching. When fringeheads find satisfactory shelters, they still need the proper currents to bring food within reach. Thus no fish exists alone — rather, each is an integral part of a community or environment, and each community, environment, or habitat must offer optimal conditions if the species is to survive.

Where the ocean bottom is rocky, kelp forests usually prevail and the substratum is lush with vegetation. These areas typically support the richest assortments of plants and animals, including fishes. One such area at Diablo Cove, San Luis Obispo County, California, was studied intensively during a two-year period, and 77 kinds of fishes were found inhabiting depths shallower than 70 feet (22 m). At three localities further south (Del Mar and La Jolla, California, and Punta Banda, Baja California), fishes were collected by surrounding some small nearshore kelp beds with a net and then using ichthyocides and explosives in the enclosed 1/2- to 1 1/2-acre (.2 to .6 hectares) area. Water depths at these three

1

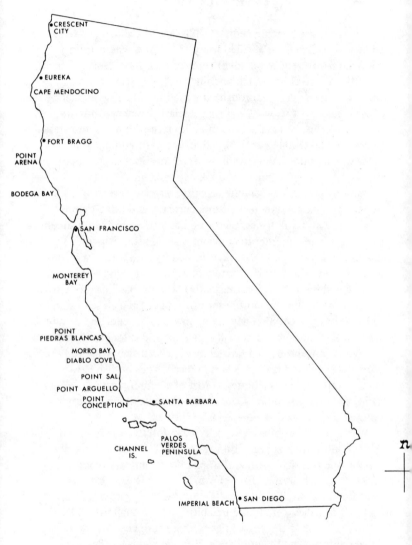

CRESCENT
CITY

● EUREKA
CAPE MENDOCINO

● FORT BRAGG
POINT
ARENA

BODEGA BAY

● SAN FRANCISCO

MONTEREY
BAY

POINT
PIEDRAS BLANCAS
MORRO BAY
DIABLO COVE
POINT SAL
POINT ARGUELLO
POINT
CONCEPTION   ● SANTA BARBARA

PALOS
VERDES
CHANNEL   PENINSULA
IS.

IMPERIAL BEACH ● SAN DIEGO

n

Coastal landmarks and localities along California and Baja
California for tidepool and nearshore fishes.

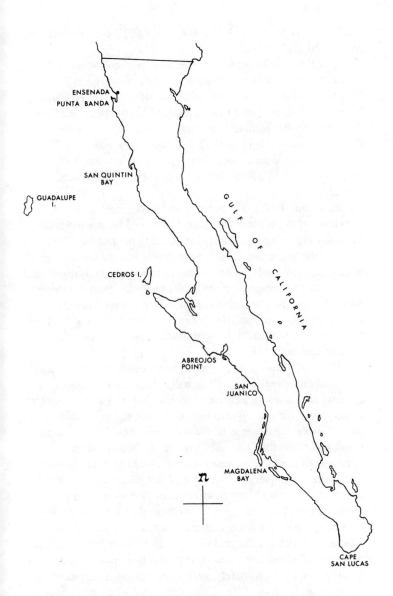

ENSENADA
PUNTA BANDA

SAN QUINTIN
BAY

GUADALUPE
I.

CEDROS I.

GULF OF CALIFORNIA

ABREOJOS
POINT

SAN
JUANICO

n

MAGDALENA
BAY

CAPE
SAN LUCAS

localities were 35, 25, and 35 feet (10.6, 7.6, 10.6 m) respectively, and 24, 26, and 39 kinds of fishes were taken.

On flat, nonrocky bottoms, several years of collecting with a beach seine on numerous stretches of sandy coast and in back bays between Point Conception and the Mexican border seldom have yielded more than 15 to 20 kinds of fishes at any one locality. When the catches from all localities, seasons, and years were lumped, however, there were 71 different kinds, including 12 species of sharks, skates, and rays — probably less than half what would be found in a rocky habitat if sampled with the same intensity.

One would think that a habitat which takes such a beating from the elements as the intertidal zone does would be indestructible, but this is not the case. In addition to being pounded by the breaking surf, the entire intertidal zone is exposed to the sun and wind during daylight hours, and to a drop in temperatures at night. Both day and night temperatures are colder in winter, than in summer. When it rains, the plants and animals which are acclimated to salt water often are inundated by fresh water. Predation alternates between terrestrial and aquatic animals, depending upon the stage of the tide.

All of these factors, in moderation, are conditions of life for intertidal inhabitants, but occasionally Mother Nature goes on a rampage that is intolerable even for these hardy creatures. Severe storm waves will tear up kelp beds and move gigantic boulders around, wreaking havoc with the area's inhabitants; kelp, clams, snails, and fishes may be tossed high on the beach and left stranded. In summer, red-tide organisms often multiply in such astronomical numbers and use up so much dissolved oxygen that the fishes and other living creatures suffocate. Prolonged runoff during the rainy season can kill nearshore animal life, including sedentary fishes. Predators and diseases can decimate animal populations in localized areas, and prolonged warm-water periods can wipe out kelp beds and cause their residents to move, starve, or be preyed upon. Fortunately, such rigors seldom affect more than a small fraction of the environment; but when man's special talent for disrupting and destroying becomes a factor, either by accident or on purpose, the natural resource invariably comes out second-best.

4

Man's Impact on the Nearshore Environment

Man's impact on the nearshore environment has become almost as varied as the commercials on TV, and even more difficult to ignore. Whether the impact is accidental or planned, direct or indirect, subtle or drastic, chronic or sporadic, localized or extensive, one can rest assured that it will be detrimental to the natural resources. To date, most attempts to control or regulate these adverse activities can be likened to locking the barn after the horse has been stolen. Man-related activities which most seriously affect the nearshore environment generally fall into three broad categories: waste disposal, habitat destruction, and resource exploitation.

Waste Disposal

Approximately 3 billion gallons (11.4 million kl) of industrial and domestic waste products pour into California's marine environment daily. These wastes travel through some 700 outfalls which either empty directly into the ocean or into a convenient gully, stream, or river — a trip which takes a bit longer but ends in the same place. At 3 billion gallons (11.4 million kl) per day, in one year a 2.8 million-acre (1.1 million hectares) lake 80 miles long by 55 wide (130 x 89.5 km), could be filled to a depth of 1 foot (.3m). If this quantity of waste were spread evenly over the entire ocean, adversities probably could not be detected, but this is not the case. The great bulk of these waste products end up in rather limited areas near the state's major population centers, and the concentrations that pour into these areas cannot be assimilated by the nearshore marine environment, where dilution and dispersal rates are minimal.

The very complexity of today's industrial technology dictates that every conceivable type of chemical or mineral element is going to be flushed down a drain and end up in the ocean. Fortunately for man, not many of these byproducts of the modern world are assimilated by species which are harvested for food, or the threat to health would be truly frightening. As it is, few will forget the widely publicized finding of DDT, mercury, lead, cadmium, and radioactive wastes in fishes and other products from

5

the sea, or the constant threat of typhoid from eating clams and mussels which are bathed in domestic sewage.

Not all waste products go directly into a sewer system, however. Each year in California, countless thousands of tons of toxic substances enter the atmosphere as air pollutants, and the bulk of these settle back to earth within a few miles of the sea. During runoff from the season's first heavy rainfall, these former air pollutants are washed from rooftops, driveways, streets, alleys, fields, and hillsides, and are on their way to the ocean. While they are airborne, these substances cause respiratory problems and eye irritations, kill terrestrial vegetation, melt women's hosiery, and corrode metals. It hardly seems necessary to spell out their impact on the nearshore resources, once they reach the sea.

Not all waste products are chemically oriented, however. Heated water can destroy kelp beds, but it attracts warm-water food and game fishes. In this case it is not a simple matter of an enhanced fishery canceling out the loss of kelp; rather, an entire community can be destroyed. And as with sewer outfalls, heated waters are not spread evenly along the entire coast. The greatest sources of thermal waste are the coastal electric generating plants which use ocean water to cool their generators. One such nuclear plant when operating at capacity, will use 2 million gallons (7.6 million l) of water per minute, 24 hours per day, 365 days per year. The effects of one such generating plant very likely will be reflected by kelp-bed losses 10 to 20 miles (16 to 32 km) away from the discharge site.

Oil in the marine environment was widely publicized as a result of a man-caused leak in the Santa Barbara Channel in February 1969, and little can be added here. What may not be known, however, is that in 1970 an estimated 3.3 million tons (2.9 million metric tons) of petroleum products (oil), arising from both natural and man-related causes, were polluting the world's oceans each year. It would be safe to assume that a substantial portion of this pollutes the waters off California.

Habitat Destruction

Habitat destruction, our second broad category, also comes in a variety of forms — many under the guise of "human progress."

Almost every year one or more of our lagoons, back bays, or estuaries, which were limited in number to begin with, is filled for highway construction or a housing tract, or dredged as a small boat harbor. Thus animal and plant communities which live only in such marsh-type habitat are sacrificed in the interest of "progress," although in these cases it can be argued that the progress is in reverse.

Equally abusive of the habitat are the countless thousands of students, tourists, and beach strollers who, when the tide is out, swarm over the shore in search of any living creature they can collect as a curio, for classroom study, or simply because it is there. Not only is the substratum denuded by this "bucket brigade," but their turning of rocks and tearing up of reefs during their search for "goodies" destroys great quantities of the uncollected plants and animals.

Habitat, especially rocky habitat, also suffers irreparable damage from the millions of tons of sediments which pour into the ocean annually through the state's multitude of submarine outfalls and storm drains.

Resource Exploitation

Our third broad category, resource exploitation, perhaps enjoys greater *direct* mass participation than do the other two. Both commercial and sport fishermen are involved in exploiting our marine resources. Fishes, shrimps, crabs, lobsters, squids, abalones, and sea urchins, to name a few, are harvested for food. Mussels, worms, shrimps, sand crabs, anchovies, and such are utilized for bait. Tons and tons of kelp are processed for use in a multitude of products including toothpaste, ice cream, pharmaceuticals, textile dyes, and food supplements. Then there is the "bucket brigade" mentioned earlier.

Unmanaged exploitation of our marine resources, or unsound management practices, can result in irreversible downward trends in populations or in entire communities. Sport fishermen constantly point to the catastrophic decline of the Pacific sardine during the 1930s and '40s as a shining example of overfishing. Many intertidal fishes guard their territories and nests without regard for the size or intent of intruders, and thus are extremely vulnerable. Because of this factor alone, it was necessary to protect the garibaldi against amateur spearfishermen who felt their status

7

was at stake unless they harpooned one of these colorful fish.

Although some resources are exploited unintentionally, the effect on the fauna or flora can be equally devastating. The thousands of pounds of fishes, crabs, squid, and other organisms drawn into steam-generating plants along with cooling waters are just as lost to the overall population as are those taken for food, bait, or curios. Perhaps even greater losses occur among eggs and larvae which are in the cooling waters, but there is no good way to measure these losses as yet.

None of man's activities which have an impact upon the marine environment goes unobserved or unchecked. Dozens of organizations and thousands of individuals are involved in monitoring, evaluating, setting standards, and otherwise protecting our resources and the environment against detrimental or potentially detrimental activities, and in fixing the blame when adversities do occur. All levels of government — international, national, state, and local — have vested interests in protecting our oceanic heritage, but the ultimate responsibility lies with each individual among us. Without our help and cooperation, little if any of the incredible variety of plants and animals living off our coast today can or will survive for future generations to enjoy.

Scope of Coverage

Our primary concern in writing this book has been to cover all families of bony fishes not previously discussed in our *Deep-Water Teleostean Fishes of California* and *Marine Food and Game Fishes of California* (California Natural History Guides: 25, 1968, and 28, 1971). Since most of the families discussed here contain forms which inhabit tidepools or shallow nearshore waters, this book is titled accordingly. Our second concern was to give a complete report on this inshore fish fauna, and to accomplish this it has been necessary to repeat in part the accounts of several food and game fishes which inhabit this environment and are the sole members of their particular families.

One or more members of each family is illustrated and discussed in the narrative reports. When a family contains more than one species, and thus afforded opportunity for a choice, we gave preference to forms which previously had received little attention, or which had some unique behavioral trait or facet of life history that we felt was worth reporting. Much of our life-history information is based upon notes we have accumulated over the years, or research we undertook to obtain vital statistics not previously available.

When there are fewer than eight or ten species in a given family, we have constructed a rather simple one- or two-character key as an aid in distinguishing them. For some families there are keys in one of our other two books in this series, so if new information was unavailable we did not duplicate these. For the larger families (e.g., Agonidae, Cottidae, Embiotocidae, Gobiidae, Pleuronectidae, Sciaenidae, Scorpaenidae, etc.), technical or popular publications are available which will aid in identification. We have included these titles in the list of helpful references.

All drawings are based upon actual specimens, and various body proportions, fin lengths, fin positions, and other anatomical features are depicted as they appear in the specimen drawn, believed to be typical for the species. In several cases we felt that a flat-out side view would not do justice to the fish in question, so these have been posed in three-quarter views. All drawings have been examined critically by several experts in fish identification.

We have presented the family accounts alphabetically (Agonidae through Trichodontidae) in order to simplify the task of locating a family account without having to check the index. At first this may help only knowledgeable professionals (ichthyologists, fishery biologists, etc. and students) but we feel that fishermen are anxious to learn the families of the fishes they see and catch, and thus will soon be able to ignore the index too.

As in our previous two volumes, we have included all family members in our checklist (Appendix I), even though some species may not inhabit shallow water at any stage of their life. The checklist too is alphabetical by family, and within each family the members are noted alphabetically by genus and species.

Common names have always been a problem, and they will continue to confuse as long as there are people and things to name. Some fishes are known by one name in southern California, by

9

another in northern California, and by still other names both north and south of the state. Other fishes are called something by sport-fishermen and something else by commercial fishermen, and then there is the fishery biologist's vernacular. To a Californian, a "rockfish" is a member of the family Scorpaenidae; to an easterner, it is a name commomly applied to striped bass. To quote from the American Fisheries Society's booklet *A List of Common and Scientific Names of Fishes from the United States and Canada*: "There is clear need for standardization and uniformity in vernacular names, not only for sport or commercial fishes, but as trade names, for aquarium fishes, in legal terminology, and as substitutes for scientific names of almost any fish in popular or scientific writing."

We concur wholeheartedly, and with few exceptions have used the names given in the AFS checklist. One notable exception is our use of "rockcod" instead of "rockfish" for members of the genus *Sebastes*. Historically, "rockcod" has been used by California's sport and commercial fishermen for the 57 kinds of *Sebastes* caught in Californian waters, and since the bookish "rockfish" recommended by scientists more than a quarter-century ago has not been picked up by industry personnel, we can see no reason to continue its use.

FAMILY ACCOUNTS

Agonidae (Poacher Family)
Rockhead
Bothragonus swanii (Steindachner, 1876)

Distinguishing characters. — This is the only poacher in our waters which has smooth body plates, two dorsal fins, and a deep pit in the top of the head. In addition, no other poacher has such a stout, distinctively shaped body.

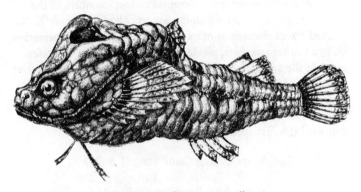

Fig. 1.   *Bothragonus swanii*

Natural history notes. — *Bothragonus swanii* ranges from Alaska to Diablo Cove, California, just south of Morro Bay. Within this area they have been collected intertidally and to depths of 60 feet (18.3 m). The largest individual on record was 3.5 inches long (89 mm), but we have no idea as to its age. A 2.4-inch (6.1 cm) rockhead that was three years old weighed one-fourteenth of an ounce (2 g); we have no information regarding age or size at maturity, spawning habits, or other facets of reproduction, however.

Their food consists primarily of tiny crustaceans (shrimp like and crab like creatures), which they pick up on the bottom or very close to it. We do not know of any specific predators, but suspect that rockheads fall prey to any carnivore which chances upon them.

Capture data. — Rockheads take on a coloring which fits in

11

well with their background (e.g., brown with dusky bars, orange with bluish bars, rusty-red with brownish bars, etc.), so they are difficult to see in their natural habitat. Skindivers who are experienced collectors capture numerous individuals for aquaria, primarily north of California, where they are most abundant at shallow depths. Mostly, however, they are not seen unless fish collections are being made by fishery biologists or university or museum personnel using ichthyocides (chemical preparations capable of suffocating fishes and some other forms of marine life that have gills).

Other family members. — Seventeen other members of the family are know from Californian waters. All can be easily distinguished by an assortment of external characters. For information on these other poachers, please refer to the checklist and references in the appendices, and to the family account in *Deep-water Teleostean Fishes of California.*

Meaning of name. — *Bothragonus:* pit *Agonus; swanii:* in honor of James G. Swan of Port Townsend, Washington, who probably collected the first specimen.

Ammodytidae (Sand Lance Family)
Pacific Sand Lance
Ammodytes hexapterus Pallas, 1814

Distinguishing characters. — The sand lance is recognizable by its elongate silvery body, long dorsal fin, deeply forked tail, and lack of pelvic fins. The lateral line, high along the back, and a longitudinal fold running the length of the fish on each side, at a level with the belly, are sufficient to clinch its identification.

Fig. 2.   *Ammodytes hexapterus*

Natural history notes. — *Ammodytes hexapterus* ranges in the eastern Pacific from the Bering Sea to Balboa Island, Orange

12

County, California; in the western Pacific, southward to Japan. They are most often found on sandy bottoms from the intertidal into depths of 60 feet (18.3 m), but we know of one trawled from 156 feet; in surface waters, schools of larvae and juveniles are often found considerable distances offshore over quite deep waters. Adults often bury themselves in the sand, a habit which has earned them part of their common name. Although reported to attain a length of 10-1/4 inches (26.4 cm) in the Bering Sea, few along the California coast reach a length of 8 inches (20.3 cm). An individual 6 inches (15.2 cm) long was four years old and weighed just over a quarter ounce (7 g). The oldest Pacific sand lance we have seen was eight years old; it was 8 inches (20.3 cm) long and weighed about 1-1/4 ounces (34 g).

Although we lack information on age at first maturity, fecundity, spawning behavior, and most other facets of reproduction, it is thought that most spawning takes place in relatively shallow water during spring months. Inch-long (2.5 cm) larvae are abundant in some areas during early summer.

Adult sand lances feed primarily on a variety of small crustaceans, including eggs of some forms, but will eat most other kinds of bite-sized food items which are found in their habitat. In turn, sand lances are preyed upon by almost every kind of fish-eating fish, bird, and mammal large enough to handle them; many are even captured and eaten by crabs.

Fossil otoliths from *Ammodytes hexapterus* have been found in several Pliocene and Pleistocene deposits between San Pedro, California, and central Oregon.

Capture data. — Pacific sand lances have been captured incidentally in a variety of seines and nets, although they seldom are used for food. In addition, they are frequently killed with ichthyocides during fish-collecting expeditions conducted by fishery biologists and university and museum personnel.

Other family members. — No other family member is known in the eastern Pacific Ocean.

Meaning of name. — *Ammodytes:* literally a sand diver, or burrower; *hexapterus:* six-winged (probably alluding to the combination of fins and skin folds).

Anarhichadidae (Wolffish Family)
Wolf-eel
Anarrhichthys ocellatus Ayres, 1855

Distinguishing characters. — The lack of pelvic fins, extremely elongate tapering body, slender and almost indistinguishable caudal fin, large conical canines and molarlike jaw teeth, and distinctive color (large ocellated black spots) will separate the wolf-eel from all other fishes in our waters. The wolf-eel is closely related to the blennies; it is not an eel.

Fig. 3. *Anarrhichthys ocellatus*

Natural history notes. — *Anarrhichthys ocellatus* ranges from Kodiak Island, Alaska, to Imperial Beach, California, but is not common south of Point Conception. In the cold northern waters they inhabit relatively shallow rocky areas, but to the south they are found at much greater depths. One taken near La Jolla was hooked in 400 feet (122 m) of water, and others have been caught off southern California in depths greater than 250 feet (76.2 m).

Various accounts report that they attain a length of 8 feet (2.4 m); but this is obviously an estimate made by some writer several decades ago, and subsequent authors have "followed the leader." The largest authenticated wolf-eel appears to be a 6-foot 8-inch (203 cm) fish speared off Rosario Beach in the San Juan Islands, Washington, in 1962; this fish weighed 40 pounds and 10 ounces (18.4 kg). Three smaller wolf-eels, 38, 51, and 61 inches (96.5, 129.5 and 154.9 cm) long, weighed 2-1/2, 7, and 12-1/3 pounds (1.1, 3.2 and 5.6 kg) respectively. Based upon these lengths and weights, an 8-footer (2.4 m) would have to weigh over 100 pounds (45.4 kg) and possibly as much as 150 pounds (68 kg). An immature male 35 inches (89 cm) long appeared to be four years old, judged by some rather vague growth rings on its otoliths.

Spawning takes place during winter months, and the whitish eggs are deposited in a mass on a protected surface of a rocky cave or crevice. The eggs adhere to the rocky substratum, and both parents remain on guard until hatching takes place. One nest contained an estimated 7,000 eggs, but information on the size of the female is lacking. We have no information on age or length at first maturity, number of eggs for a female of any given size, time required to hatch, or other facets of reproduction.

The wolf-eel stomachs we have examined have contained a preponderance of crab remains. Other items we have observed were sea urchin fragments, small snails, including abalones, and an occasional piece of fish. There is a report of a 16-1/2-inch (41.9 cm) wolf-eel having been found in a salmon stomach, but we do not know of any other predation.

Wolf-eel teeth have been found at an Indian village site at Diablo Cove (San Luis Obispo County), the fish having been harvested and eaten by the inhabitants over a period of at least 9,000 years.

Capture data. — Sportfishermen take an estimated 200 wolf-eels a year, according to a survey conducted from 1957 to 1961. About two-thirds of these are speared by skindivers, while most of the remainder are caught by skiff fishermen. Those taken on hook and line usually are attracted by an anchovy, some other small bait fish, or a piece of abalone.

There is no commercial fishery for wolf-eels.

Other family members. — No other member of the family is known in the eastern Pacific Ocean.

Meaning of name. — *Anarrhichthys: Anarhichas* fish, for its resemblance to an Atlantic relative; *ocellatus:* with eyelike spots.

Antennariidae (Frogfish Family)
Roughjaw Frogfish
Antennarius avalonis Jordan and Starks, 1907

Distinguishing characters. — The stalked, fanlike pectoral fin and a small gill opening at approximately the "armpit" are sufficient to distinguish the roughjaw frogfish from all other Californian fishes. The bizarre shape (outline) of this loose-skinned frogfish is also distinctive.

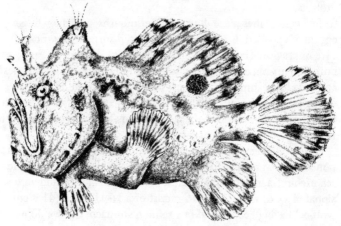

Fig. 4. *Antennarius avalonis*

Natural history notes. — *Antennarius avalonis* was first described from a specimen taken at Avalon, Santa Catalina Island, California, but is now known to range from there to Peru and offshore at the Galapagos Islands. They are generally sedentary in nature, and have been seen or captured on the bottom from the intertidal zone into depths of 360 feet (109.7 m). The one taken at Avalon was 13-1/2 inches (34.3 cm) long and is the largest

16

known; most individuals we have seen were less than 5 inches (12.7 cm) long. We have no information regarding ages or reproductive habits, but frogfishes in general are said to liberate their eggs in a single long ribbonlike mass which floats like a raft of logs on the way to a sawmill. This gelatinous mass is enormous in proportion to the fish that deposits it.

The roughjaw frogfish is almost entirely carnivorous and extremely voracious. Younger individuals feed rather heavily on crustaceans, but as they grow larger they consume more and more fish, sometimes catching and eating fishes as long as themselves. In an aquarium, and probably in nature, they usually spend their time sitting on the bottom waiting for food to swim by, and then wave or twitch their lures to attract it within reach. Frogfishes in turn are occasionally eaten by larger predators which live in the same environment, especially other fishes.

Capture data. — Frogfishes sometimes are caught on hook and line, and we know of one that even bit on a piece of angleworm, but most are taken by biologists and other individuals making collections in rocky habitat. Sometimes they are found in large masses of seaweed which are torn loose from the bottom and hauled aboard ship with anchors, dredges, nets, and similar devices.

Other family members. — No other frogfish is known from Californian waters or within several hundred miles to the south, at least.

Meaning of name. — *Antennarius:* derived from antenna, referring to the fishing lure; *avalonis:* for Avalon Bay, the locality of first capture.

Apogonidae (Cardinalfish Family)
Guadalupe Cardinalfish
Apogon guadalupensis (Osburn and Nichols, 1916)

Distinguishing characters. — The two high but short-based dorsal fins, very long caudal peduncle, and bright body color of this little fish will distinguish it from any other species in Californian marine waters. It is bluish-gray above and reddish-orange on the lower sides and belly.

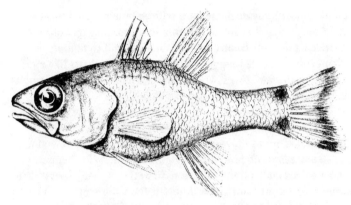

Fig. 5.   *Apogon guadalupensis*

Natural history notes. — *Apogon guadalupensis* has been found in our waters only at San Clemente Island; primarily it is found 214 miles (344 km) south of there at Guadalupe Island and along the mainland coast from Magdalena Bay to and throughout the Gulf of California. During the daytime they hide in rocky crevices and caves at depths to 60 feet (18.3 m) or more, but at night they move off the bottom and out into open areas, where they feed until the approach of dawn. They feed primarily on planktonic crustaceans and small fishes.

We have no information on reproductive behavior, spawning seasons, age at first maturity, and such, but at maximum size of about 5 inches (12.7 cm) they appear to be five or six years old, judged by the number of growth zones on their otoliths. Individuals 2-1/2 to 3 inches (6.4 to 7.6 cm) long are two years old.

A cardinalfish otolith, probably from this species, was found in a Pleistocene deposit at Guadalupe Island.

Capture data. — Cardinalfishes seldom are captured except by scientific personnel using ichthyocides or fine-mesh gill nets. Scuba divers sometimes catch small individuals alive by using slurp guns or fine-mesh handheld bag nets. Cardinalfishes are also fairly easy to trap on occasion.

Other family members. — The Guadalupe cardinalfish is the only member of the family known from nearshore waters off California.

Meaning of name. — *Apogon:* without a beard, to distinguish it from a "look-alike" which has fleshy barbels under the chin; *guadalupensis:* for Guadalupe Island, Mexico, the locality of first capture.

Atherinidae (Silverside Family)
Topsmelt
Atherinops affinis (Ayres, 1860)

Distinguishing characters. — The typical body shape, large scales, silvery bar on the side, and tiny first dorsal fin which is well separated from the second dorsal will distinguish the three members of the silverside family from other marine fishes. Topsmelt have forked jaw teeth (a magnifying glass must be used to see this character) which will set them apart from grunion, which lack jaw teeth, and jacksmelt, which have simple conical teeth.

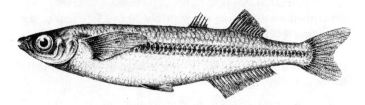

Fig. 6.   *Atherinops affinis*

Natural history notes. — *Atherinops affinis* ranges from 4 miles (6.4 km) west of Sooke Harbour, Vancouver Island, to and into the Gulf of California. Within this range, and depending upon habitat (back bays, kelp beds, offshore islands) and geographic locality (extreme south), the topsmelt populations are separated into several subspecies. Generally they are found in loose schools at or near the surface of the water. The largest reported individual was just under 14-1/2 inches (36.8 cm) long and weighed 11 ounces (310 g). We have no idea as to the age of a topsmelt that size, but a 12-inch (30.5 cm) fish was four years old, and several about a half-inch (1.3 cm) longer were five.

Some topsmelt will spawn when two years old, and most will spawn when three. Spawning takes place primarily in the

19

late winter and spring, and the relatively large eggs adhere to giant kelp and other plant life. Topsmelt typically feed within 10 to 15 feet (3 to 4.6 m) of the surface over shallow rocky areas or in kelp beds. Adults eat tiny planktonic crustaceans almost exclusively, while quantities of algae and kelpfly larvae have been found in stomachs of juveniles taken close to shore and in tidepools. Topsmelt are fed upon by a variety of seabirds and predatory fishes, but seldom in quantity by a single predator.

Topsmelt otoliths have been found in several Pliocene and Pleistocene deposits in southern California, and in a few Indian middens.

Capture data. — Topsmelt are caught both by pier fishermen using hook and line and by commercial fishermen using lamparas, or similar encircling nets. They seldom are distinguished from jacksmelt in catch statistics, and the combined commercial catch for these seldom exceeds 500,000 pounds (226,750 kg) per year. An estimated 320,000 jacksmelt and topsmelt are harvested by sportfishermen each year, and probably 20 to 25 percent of these are topsmelt which average 1/4 pound (112 g) each.

Other family members. — Three kinds of silversides inhabit our waters. These can be easily distinguished by the following simple key:

1. If the cheek area is deep blue in color and there are no teeth in the jaws, it is a California grunion.
2. If the cheek area is bright lemon-yellow, and
   a. if the jaw teeth are simple and in several rows, it is a jacksmelt.
   b. if the jaw teeth are forked at their tips and in a single row it is a topsmelt.

Meaning of name. — *Atherinops:* like *Atherina,* a related genus; *affinis:* related — presumably to associated species.

Aulorhynchidae (Tube-snout Family)
Tube-snout
Aulorhynchus flavidus Gill, 1861

Distinguishing characters. — The pencil-sized body, tubular snout with tiny terminal mouth, and 23 to 26 short spines in

20

the first dorsal fin will distinguish the tube-snout from all other fishes in our waters.

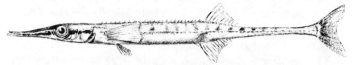

Fig. 7.   *Aulorhynchus flavidus*

Natural history notes. — *Aulorhynchus flavidus* ranges from Sitka, Alaska, to Rompiente Point, Baja California. They usually are found in loose schools among beds of giant kelp at depths to and exceeding 100 feet (30.5 m) but on one occasion a dense school containing millions of individuals was encountered over deep water south of Santa Rosa Island. The maximum recorded length is 7 inches (17.7 cm) for a ripe female taken in July, near Morro Bay. This fish weighed about one-third of an ounce (9 g) and contained slightly more than 400 eggs. Several tube-snouts which were 5-1/2 to 6 inches (13.9 to 15.2 cm) long appeared to be five years old, judged by some rather vague growth zones on their otoliths, and the 7-inch female appeared to be nine. In British Columbia the young are thought to reach 2 inches (5.1 cm) in two months, and some fishery biologists believe that tube-snouts mature, spawn, and die in one year. Obviously, additional studies are needed on tube-snout ages.

Tube-snouts are nest-builders, and their reproductive behavior has been observed on many occasions. The male typically selects a nesting site in giant kelp beds at depths below 25 feet (7.6 m) and binds some of the seaweed fronds together with sticky web-like strands which are extruded from the urogenital area. Females then deposit clusters of eggs around the kelp stipe, slightly above the prepared nest, and the male after fertilizing the eggs remains on guard until they hatch. Usually several such clusters of eggs will be deposited at a given nest by different females; sometimes as many as ten clusters can be found on a single stipe. Intruders into the area are driven away by the guarding male. Considering the size of the tube-snout, the eggs are quite large — about one-twelfth inch (2 mm) in diameter. Nests with eggs have been noted as deep as 125 feet (38 m) during all months from February through July.

21

Adult tube-snouts feed primarily upon planktonic crustaceans, but they will also eat larval fish, tiny polychaete worms, and other food items which are small enough. In turn, their remains have been found in the stomachs of kelp bass, barred sand bass, rockcod (several kinds), lingcod, and other predatory species which inhabit the same environment.

Capture data. — Tube-snouts sometimes are captured alive for aquaria, and many collections have been made with ichthyocides for university and museum collections, but they are not harvested nor utilized by commercial and sportfishermen.

Other family members. — No other member of the tube-snout family is known from the eastern Pacific Ocean, although some scientists consider the tube-snout to belong to the stickleback family (Gasterosteidae), which contains two fresh- and brackish-water inhabitants in our state.

Meaning of name. — *Aulorhynchus:* tube snout; *flavidus:* yellow.

Balistidae (Triggerfish Family)
Finescale Triggerfish
Balistes polylepis Steindachner, 1876

Distinguishing characters. — The tough leathery skin covered with small, solidly affixed scales; the tiny gill opening above and in front of the pectoral fin; the small terminal mouth filled with close-set (rabbitlike) teeth with moderately strong cusps; and the three-spined first doral fin, which can be locked into an erect position are more than sufficient characters to identify the finescale triggerfish.

Natural history notes. — *Balistes polylepis* is endemic in the eastern Pacific, and has been captured between Crescent City, California (once), and Afuera, Peru. They are abundant throughout the Gulf of California and south through tropical waters, but their occurrence off California is sporadic and unpredictable. The early juveniles are pelagic (living at the surface and well offshore), but the adults prefer inshore areas where the bottom is sandy, or where sandy and rocky habitat intermix. They do not shy away from rocky bottom areas, however.

22

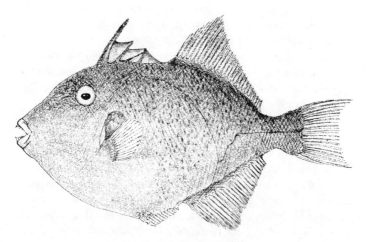

Fig. 8. *Balistes polylepis*

We have seen finescale triggerfish that were about 30 inches (76 cm) long and weighed in excess of 10 pounds (4.5 kg), and they have been reported in the literature as weighing up to 16 pounds (7.3 kg), but vital statistics apparently have never been taken on the very largest individuals. An 18-inch (45.7 cm) fish taken off Paradise Cove weighed 3-1/2 pounds (1.6 kg), but we have no information on the age or sex of this individual.

As common as this fish is in the Gulf of California, its life history has not been studied. There is no information on its age, growth, maturity, fecundity, or behavior. The few observations on its food habits are casual at best. Triggerfishes have strong jaws and teeth and are noted for their ability to bite off bits of coral and other sedentary animals that are encased in various kinds of "armor". We have found remains (shell fragments) of barnacles, clams, and snails in the stomachs of finescale triggerfish, as well as fish flesh. This is the only fish we know of that appears to relish shark meat (aside from a larger shark). In the southern Gulf of California, swarms of finescale triggerfish will be chewing on any shark carcass that is thrown overboard before it has time to reach bottom. They continue "attacking" the carcass until nothing remains.

Pelagic juveniles are fed upon by tropical tunas, wahoos, and other large predators that inhabit offshore waters, but we do not know of anything that preys on large adults.

Triggerfish teeth, possibly those of *B. polylepis,* have been found in Miocene fossil deposits near Bakersfield and Santa Ana that may be 15 to 25 million years old. Juvenile triggerfish adjust quite well to life in saltwater aquaria and are interesting to observe, but they will nibble fins and other anatomical parts from their fellow aquarium dwellers.

Capture data. — There is no sportfishery for finescale triggerfish in California. The twenty or more individuals that have been reported from our waters have been hooked, speared, or netted. Only in the case of those that were speared did the fisherman make an effort to "catch" the species. Finescale triggerfish are excellent table fare, but they must be skinned, and because of their tough hide many fishermen become discouraged and discard them. In many parts of the world, species of triggerfish are extremely toxic to humans and can cause an agonizing death if eaten, even in small amounts. The finescale triggerfish has never been found toxic, however.

Other family members. — One other triggerfish has been reported from California, but only once. It is easy to identify these two triggerfish by color and body shape, but the characters noted below are easily observed (and explained):

1. If there are 4 to 6 conspicuous longitudinal grooves on the cheek below the eye, it is a redtail triggerfish.
2. If there are no grooves on the cheek, it is a finescale triggerfish.

Meaning of name. — *Balistes:* alluding to the trigger mechanism that comprises the spinous dorsal; *polylepis:* many scales.

Bathymasteridae (Ronquil Family)
Stripefin Ronquil
Rathbunella hypoplecta (Gilbert, 1890)

Distinguishing characters. — The blennylike body, long dorsal and anal fins composed entirely of soft rays, rounded caudal fin,

24

and straight lateral line will distinguish ronquils from all other fishes in our waters. The 12 orbital pores, a longitudinal blue stripe that runs the length of the anal fin, and the lack of long canines in the sides of the jaws will distinguish the stripefin ronquil from all other ronquils.

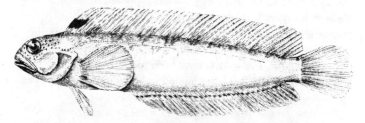

Fig. 9.   *Rathbunella hypoplecta*

Natural history notes. – *Rathbunella hypoplecta* ranges from about San Francisco Bay to the vicinity of Punta Banda, Baja California. Inhabiting rocky areas along the outer coast at depths of 30 to 300 feet (9 to 91 m) at least. They have been reported to attain a length of 8-1/2 inches (21.6 cm), but this size may apply to one of the other species of *Rathbunella* rather than *R. hypoplecta.* The largest we have seen was a 6-1/4-inch-long (15.9 cm) male that weighed 1 ounce (28 g) and was five years old. Three-year-old fish range from 4 to 5-1/4 (10.2 to 13.3 cm) inches long and average about one-half ounce (14 g) each.

Although there are several color differences between males and females, the most striking of these is the color of the tips of the anal fin rays. In males the anal ray tips are black; in females, they are white. Both sexes apparently are mature when two years old, and spawning takes place from May until October. In an aquarium, several stripefin ronquils were observed for about two-and-a-half years. The first year a single female spawned six times, at two-week intervals. Within a day or two of the time one batch of eggs would hatch (thirteen to fifteen days), she would spawn another batch. The eggs would be deposited in a secluded spot, in this case a jar in the aquarium bottom, and the male would remain on guard. An estimated 10,000 eggs were spawned each time, and one such cluster weighed nearly one-third of an ounce (8.3 g). During her second year in the aquarium, this same female spawned only three times between July and September.

Stripefin ronquils feed upon a variety of invertebrates, and apparently prefer crustaceans, but several stomachs examined at Diablo Cove contained only nudibranchs. The nest-guarding male frequently ate some of the eggs he was supposed to be protecting. We have no specific information regarding ronquil predators, but assume that ronquils would be fair game for large carnivores in their environment, especially other fishes.

Otoliths of *R. hypoplecta* have been found in a Pliocene deposit near San Pedro, California, estimated to be around 4 million years old.

Capture data. — A few ronquils have been caught by sportfishermen using tiny hooks while fishing on the bottom; others have been taken incidentally by commercial trawl fisheries. Most ronquils, however, are taken by biologists and others who are making scientific collections. They are not difficult to capture alive if one has scuba gear, and they become acclimated to aquarium life very quickly.

Other family members. — At least three members of the family inhabit Californian waters, but none is easy to distinguish without a critical examination. One of these, the northern ronquil, has 15 orbital pores, whereas the other two have 12. In another, the bluebanded ronquil, there is a blue bar behind each anal ray, while in *R. hypoplecta* a horizontal blue bar runs the length of the anal fin.

Meaning of name. — *Rathbunella:* for Mr. Richard Rathbun, then Chief of the Division of Scientific Inquiry in the U.S. Fish Commission; *hypoplecta:* folded underneath (presumably alluding to the branchiostegal membrane).

Batrachoididae (Toadfish Family)
Specklefin Midshipman
Porichthys myriaster Hubbs and Schultz, 1939

Distinguishing characters. — The scaleless body, purplish-bronze above and yellowish below, the large broad head with widely separated protrusible eyes, and the numerous rows of luminescing organs on the underside are sufficient to distinguish a midshipman from all other fishes in our waters. The specklefin midshipman is characterized by its spotted dorsal, anal, and pectoral fins, and by the second row of photophores on the throat, which form a forward-directed U.

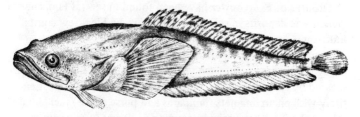

Fig. 10.   *Porichthys myriaster*

Natural history notes. – *Porichthys myriaster* ranges from Point Conception, California, to Magdalena Bay, Baja California. They are found primarily in rocky areas from the intertidal zone into depths of 414 feet (126.2 m). When in muddy areas, they bury themselves during daylight hours, but emerge at night in search of food. Males attain a considerably larger size than females, and the longest on record was an 18-3/4-inch (47.6 cm) male weighing 2.6 pounds (1187 g); a male 1-1/2 inches (38.1 mm) shorter, however, weighed 1-1/2 ounces (42 g) more than the record fish.

During April, May, and June, specklefin midshipmen move into shallow water, including bays and estuaries, where they seek out or dig nesting areas in rocky areas. The female attaches her eggs, typically 200 to 400 or more, to the underside of the nesting rock, and both the male and female stay on guard for a day or two before the female takes off. The eggs require a month or more to hatch, and during this period the male, who does not feed, becomes increasingly more emaciated and disease-ridden. Many males become so weakened by the ordeal that they die either before or after the eggs hatch.

We have no idea as to maximum age, but an 18-inch (45.7 cm) male that weighed just over 2 pounds (906 g) was eight years old, judged by some excellent growth zones on its otoliths.

Midshipmen will eat just about any type of food they can get their mouth around, whether it is alive or dead. Shrimplike crustaceans, octopi and squids, and small fish appear to make up the bulk of their diet, but many other kinds of food items have been noted in their stomachs. Specklefin midshipmen have been found in the stomachs of giant sea bass, sea lions, and porpoises, to name a few.

Otoliths of *P. myriaster* have been found in several Pliocene and Pleistocene deposits in southern California, and in a few coastal Indian middens.

Capture data. — There is no sportfishery for midshipmen, although some are caught each year on hook and line and others are speared by skindivers. The commercial catch is made almost entirely with encircling nets (lamparas and purse seines) incidental to other fisheries. They make interesting additions to aquaria; and because they live so well in captivity, they frequently are used in behavioral and other scientific studies.

Other family members. — One other midshipman inhabits our waters, but the two species are easily identified:

1. If the dorsal, anal, and pectoral fins are heavily spotted, it is a specklefin midshipman.
2. If these fins are without spots, it is a plainfin midshipman.

Meaning of name. — *Porichthys:* pore fish, with reference to the well-developed mucous system; *myriaster:* numerous stars, for the abundant photophores.

Belonidae (Needlefish Family)
California Needlefish
Strongylura exilis (Girard, 1854)

Distinguishing characters. — The greatly elongated jaws which are filled with needle-sharp teeth are sufficient to distinguish a California needlefish from all other species in our waters. If confirmation is needed, other distinctive characters include dorsal and anal fins set far back; the very long, almost round body; and the body color — greenish above and grading to silver-white below.

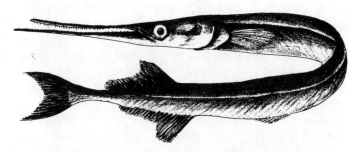

Fig. 11. *Strongylura exilis*

Natural history notes. – *Strongylura exilis* has been recorded from San Francisco to Peru, but there are only a couple of occurrences north of Santa Barbara. When seen in our waters, they usually are in small "schools" (perhaps five to fifteen individuals) at or near the surface in bays, sloughs, or harbors; they are not a fish of the open ocean. They are said to reach a length of 3 feet (91 cm), and probably do, but the largest we have measured was a 30-inch (76.2 cm) female that weighed just under 1 pound (453 g). The otoliths of this fish caught at Redondo Beach, California, in September 1960, had five winter rings, indicating an age of five years. A 19-1/2-inch (49.5 cm) male taken in March 1960 at Redondo Beach weighed less than 4 ounces (113 g) and was only one year old.

Spawning apparently takes place in the spring months, but we have no information on where they spawn, age at first maturity, number of eggs spawned each season by a single fish, or other details of reproduction. One interesting feature of the California needlefish is that only one gonad (the right one) is ever developed or functional. There is no ready explanation for this phenomenon.

Many needlefish are heavily parasitized, both internally and externally. One caught off Carpinteria, California, in September 1965 had 6 plumelike copepods (*Pennella*) sticking out of its sides. The other ends of these hairy-looking parasites were firmly and permanently attached to large blood vessels inside the fish.

Small fishes, usually anchovies, are the only food items we

29

have found in their stomachs. We know of no predators on the California needlefish, and there is no fossil record of the species.

Capture data. – The California needlefish is neither a commercial nor a game fish, although a 30-inch (76.2 cm) fish on very light tackle will put on quite an aerial display. They have been caught accidentally (or incidentally) during every month of the year, but not necessarily during any given year. Several decades ago dozens of them were speared each year from the bridge that crossed Mission Bay (San Diego) between Ocean Beach and Mission Beach. Since the bridge was removed, most catches have been made with hook and line, or in roundhaul (bait) nets. During the winter of 1972-73, quite a few needlefish were drawn into the cooling-water systems of electric generating plants along the coast between San Diego and El Segundo, California. Since these plants represent a dead end to any organism which is entrained, these needlefishes were eventually killed.

Other family members. No other member of the family is known from closer than the Gulf of California.

Meaning of name. – *Strongylura:* round tail; *exilis:* slender.

Blenniidae (Combtooth Blenny Family)
Notchbrow Blenny
Hypsoblennius gilberti (Jordan, 1882)

Distinguishing characters. – The blennylike body with a prominent fleshy cirrus above each eye, gill membranes attached to the isthmus, steep anterior profile above the low-placed mouth, and elongate dorsal fin with more soft rays than spines will distinguish *Hypsoblennius* from all other marine fishes. The V-shaped indentation in the profile of the head (side view), just above and behind the eye, will distinguish the notchbrow blenny.

Natural history notes. – *Hypsoblennius gilberti* ranges from Point Conception, California, to Magdalena Bay, Baja California. They are most abundant in the intertidal and shallow subtidal zones, but have been found as deep as 33 feet (10 m). The maximum recorded length is 5-1/2 inches (13.9 cm), and at that size they appear to be eight to ten years old.

The adults breed from about June to August, and the egg clusters are laid in nests which are guarded by the males. Each female will spawn between 600 and 1800 eggs per "clutch," and some will spawn up to three times each season. Sometimes more than one female will deposit eggs in a single nest. Once the eggs are fertilized, it takes between 5 and 18 days for them to hatch, depending upon water temperatures. The newly hatched larvae drift with prevailing currents for two or three months before settling in a suitable environment.

Notchbrow blennies are most abundant around hard sub-stratum within the intertidal zone, and within this environment they prefer living around the bases of projecting rocks, tubeworm colonies, and similar relief. Their teeth are comblike but firmly fixed, and are used in scraping or clipping. They feed heavily on limpets, tiny crustaceans, and algae, but will accept an assortment of tiny benthic invertebrates. Notchbrow blennies are preyed upon by several carnivorous fishes which live in or frequent their environment.

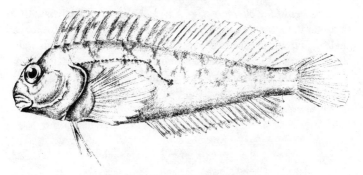

Fig. 12. *Hypsoblennius gilberti*

Capture data. — Aside from individuals captured alive for use in an aquarium, notchbrow blennies generally are not taken except by fishery biologists or museum and university personnel for fish collections or special studies. They can be caught on hook and line if both hook and bait are small enough. They adapt extremely well to life in an aquarium and are quite hardy.

Other family members. – Two other combtooth blennies inhabit our waters, but all can be distinguished by careful examination:

1. If from a side view there is a V-shaped indentation in the profile of the head, near the rear border of the eye, it is a notchbrow blenny.
2. If the anterior profile of the head is without a notch, and
   a. if the orbital cirrus is divided into filaments at its tip, it is a mussel blenny.
   b. if the orbital cirrus is not divided into filaments but is serrated along its posterior edge, it is a bay blenny.

Meaning of name. – *Hypsoblennius:* high *Blennius,* in allusion to the high front profile and resemblance to genus *Blennius; gilberti:* for Charles Henry Gilbert, renowned early ichthyologist.

Bothidae (Lefteye Flounder Family)
Speckled Sanddab
Citharichthys stigmaeus (Jordan and Gilbert, 1882)

Distinguishing characters. – Among Californian flatfish with eyes on the left side of the head, only sanddabs have a distinct caudal fin and normal scales covering the body, and lack a high arch in the lateral line above the pectoral fin. The speckled sanddab has a pectoral fin which is shorter than the head, only 8 to 10 gill rakers on the lower limb of the first gill arch, and a lower eye which is about equal to or shorter than the snout.

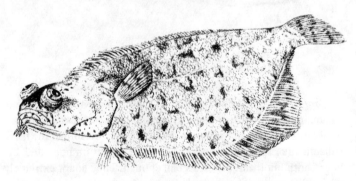

Fig. 13.   *Citharichthys stigmaeus*

Natural history notes. – *Citharichthys stigmaeus* ranges from Montague Island, Alaska, to Magdalena Bay, Baja California. Although it has been reported from depths between 10 and 1200 feet (3-365 m), it is seldom taken much deeper than about 200 (60 m). The maximum reported size is 6.7 inches (17 cm), but few individuals which are longer than 5 (12.7 cm) are seen. A 5-inch long female weighed slightly less than 1 ounce (22 g) and appeared to be two years old. Efforts to determine ages of speckled sanddabs have not been satisfactory, but it appears that they mature after their first year, and maximum age may be four years.

Spawning takes place from March through September, at least, and although many larvae are captured in plankton nets a great distance offshore, we are not convinced that the presence of these in plankton is normal. We suspect that many of these larvae have been carried so far offshore by currents that they are unable to return, and thus are lost to the population.

Stomachs of speckled sanddabs have contained a preponderance of tiny crustaceans, but polychaete worms are eaten whenever available, as is an occasional fish. In turn, speckled sanddabs are eaten by cormorants, seals, sea lions, other fishes, and crabs.

Sanddab otoliths have been found in Miocene deposits near Bakersfield, but they do not appear to be from speckled sanddabs. Otoliths of *C. stigmaeus* are abundant in many Pliocene and Pleistocene deposits throughout California, and were found once in an Indian midden at Ventura.

Capture data. As previously noted, numerous larvae are captured in plankton nets, but adults are not taken except incidental to other fisheries, especially those which use fine-mesh trawling gear. Sanddabs were the most abundant species of 104 kinds of fishes trawled in Santa Monica Bay during a 5-year study on the effects of pollution; the largest of more than 30,000 individuals taken during this study was only 5 inches (12.7 cm) long. They are easy to capture alive with handheld nets when scuba diving, and make a hardy and interesting addition to a saltwater aquarium.

Other family members. – Five other lefteye flounders are known from Californian waters, all being easily distinguished. For information on these, please refer to the checklist and the references in the appendices and to our accounts in *Marine Food and Game Fishes of California.*

33

Meaning of name. — *Citharichthys: Citharus* fish, for its resemblance to the related genus *Citharus; stigmaeus:* speckled.

Brotulidae (Brotula Family)
Purple Brotula
Oligopus diagrammus (Heller and Snodgrass, 1903)

Distinguishing characters. — The typical body shape, pelvic fins which are reduced to whiskerlike filaments, lack of a distinct caudal fin, purplish-colored body, and two overlapping lateral lines which lie parallel, one below the other, on each side, are sufficient to distinguish the purple brotula from all other fishes in our waters.

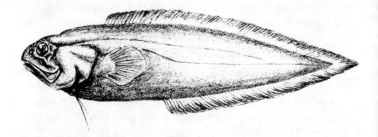

Fig. 14. *Oligopus diagrammus*

Natural history notes. — *Oligopus diagrammus* ranges from San Clemente Island, California, to Panama and the Galapagos Islands. In California, it occurs only at San Clemente Island. The species has been taken at depths of 18 to 60 feet (5.5 to 18 m), but additional specimens very likely will be found both shallower and deeper. Maximum recorded size, to date, is 8 inches (20.3 cm). They probably live at least ten years and probably longer, judged from a 5-inch (12.7 cm) specimen which was five years old.

They are quite secretive in their habits, living in rocky areas where they hide in dark caverns and crevices during daylight hours. As with several close relatives, and based upon aquarium observations, they probably move out into the open at night to feed. The only food items noted in the one stomach examined were crustaceans, although purple brotulas in an aquarium accepted the flesh of clams, shrimp, squid, and fish.

No additional details of life history are known, and there is no fossil record for the purple brotula.

Capture data. – Purple brotulas probably could be caught on hook and line, but all known specimens to date have been taken with ichthyocides during scientific collecting trips. They acclimate to aquarium life quite rapidly and are extremely interesting to observe in motion.

Other family members. – Some ichthyologists include the brotulas in the cusk-eel family, because they believe there are no really clear-cut or definitive characters for separating all members of the two families, but for representatives occurring off our coast we prefer to retain the brotulas in their own family. To date five species belonging to five genera have been reported from Californian waters:

1. If there are no pelvic fins, it is a paperbone brotula.
2. If pelvic fins are present and the lowermost pectoral rays are filamentous and separate from the uppermost rays, it is a roughhead brotula.
3. If pelvic fins are present, the pectorals are "normal," and
   a. there is a distinct caudal fin, it is a red brotula.
   b. the caudal fin is continuous with the dorsal and anal fins, and
      i. there is one lateral line, it is a rubynose brotula.
      ii. There are two overlapping lateral lines, it is a purple brotula.

Meaning of name. – *Oligopus:* small foot, in allusion to the pelvic fins; *diagrammus:* figure, probably in reference to the "double" lateral line.

Chaetodontidae (Butterflyfish Family)
Scythe Butterflyfish
Chaetodon falcifer Hubbs and Rechnitzer, 1958

Distinguishing characters. – The typical butterflyfish body shape and rather acutely tapered snout, in conjunction with the dark scythe-shaped mark on the side of this fish, are sufficient to distinguish it from any other species in our waters.

35

Fig. 15. *Chaetodon falcifer*

Natural history notes. – *Chaetodon falcifer* has been recorded from Santa Catalina Island and La Jolla, California, to the Galapagos Islands. It is one of several species of deep-living butterflyfishes known throughout the world, having been recorded as shallow as 40 feet (12.2 m) and as deep as 490 (149 m). Maximum size observed to date is about 6 inches (15.2 cm).

Because it is secretive in its behavior, and lives at such relatively great depths, few individuals have been collected and literally nothing is known of its life history. The long tapering snout is well suited for probing into crevices and other hard-to-reach places. Close relatives are known to pinch off and eat the tube feet of sea urchins and other echinoderms, and also feed upon crustaceans, worms, and other small invertebrates. There is no reason to suspect that the food habits of *C. falcifer* are any different.

Capture data. To date the scythe butterflyfish has been taken only by scientific personnel who were making fish collections while using scuba gear. Some have been speared and others have been killed with ichthyocides. They also have been observed in their natural habitat by personnel in a submersible or "diving saucer." They should make excellent additions to aquaria.

Other family members. — One other butterflyfish has been collected in our waters, at San Diego over 100 years ago. The two are easily distinguished:

1. If there is a scythe-shaped dark bar on the side of the fish, it is a scythe butterflyfish.
2. If there are vertical dark bars on the side of the fish, it is a threebanded butterflyfish.

Meaning of name. — *Chaetodon:* bristle tooth; *falcifer:* scythe-bearer.

Clinidae (Kelpfish Family)
Spotted Kelpfish
Gibbonsia elegans (Cooper, 1864)

Distinguishing characters. — The typical kelpfish body shape, pointed snout, relatively tiny mouth, rounded caudal fin, and long dorsal fin, which is slightly elevated both anteriorly and posteriorly and contains more spines than soft rays, will distinguish

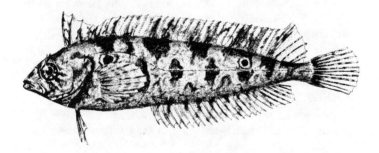

Fig. 16.   *Gibbonsia elegans*

kelpfish (*Gibbonsia*) from other marine fishes. The spotted kelpfish mimics habitat coloration and may be green, light brown, or
reddish with lighter spotting. The unequal spacing of the pos-
teriormost dorsal rays and scalation which extends onto the caudal
fin will distinguish *G. elegans* from other members of the genus.

Natural history notes. – *Gibbonsia elegans* ranges from Point
Piedras Blancas, California, to Magdalena Bay, Baja California. It
has been found at all depths from the intertidal zone to 185 feet
(56.4 m), but is not common below about 50 feet (15 m). The
maximum recorded size is 6.2 inches (15.7 cm) and about 1 ounce
(28 g) in weight; a 4-1/3-inch (10.7 cm) female weighed slightly
less than one-half ounce (12 g). Spotted kelpfish first spawn at age
two, and appear to live for at least seven years.

Spawning takes place from January to June, at least, and the
eggs are laid in a cluster or nest in short seaweeds growing on the
bottom. The white egg mass is about 1 inch (2.5 cm) in diameter,
and after spawning the male remains on guard until they hatch.
A female will spawn more than once each season.

Spotted kelpfish feed upon a variety of tiny invertebrates, especially crustaceans (isopods, amphipods, and crabs), mollusks
(gastropods), and polychaete worms. They also consume fair
quantities of algae. Remains of *Gibbonsia* have been found in
stomachs of bass and other predators which live in the same
environment.

Capture data. – Very few spotted kelpfish are taken on hook
and line, but such captures are not unknown. They are easy to
catch alive while using scuba gear, and many are taken for home
aquaria. For the most part, however, kelpfish are collected by
biologists, museum personnel, and others who are using icthyo-
cides in subtidal rocky habitat.

Other family members. – For information on other family
members, please refer to the checklist and references in the appendices.

Meaning of name. – *Gibbonsia:* in honor of Dr. William P.
Gibbons, one of the early naturalist at the California Academy
of Sciences; *elegans:* elegant.

Onespot Fringehead
Neoclinus uninotatus Hubbs, 1953

Distinguishing characters. — The blennylike body, very large mouth containing conical teeth, elongate dorsal fin with more spines than soft rays, gill membranes attached to the isthmus, and quite long fleshy cirrus over the eye will distinguish fringeheads from other Californian marine fishes. The single large blueblack ocellus bordered by yellow, which lies between the first and second spines of the dorsal fin, will distinguish the onespot fringehead from its relatives.

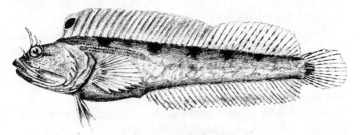

Fig. 17.   *Neoclinus uninotatus*

Natural history notes. — *Neoclinus uninotatus* ranges from Bodega Bay, California, to Ensenada, Baja California. Within this range they inhabit bottles, cans, old shoes, tires, pieces of pipe, and other sheltered "homes," either natural or man-made. They prefer depths between about 10 and 90 feet (3 to 27 m), and rest in their selected home with just the mouth and head exposed. Intruders, whether man or fish, are greeted by a large gaping mouth, and if this doesn't drive the unwanted visitor away, the fringehead will often "charge" while snapping its jaws rather vigorously. Males appear to grow larger than females. An 8-inch (20.3 cm) male weighing just over 2 ounces (65 g) appeared to be only three years old, while the largest one seen to date, at 9-3/4 inches (24.8 cm) and just over 6 ounces (175 g), was seven years old.

39

Spawning takes place during April, May, and June, at least, and a single female may spawn more than once each season. The orange-colored mass of eggs is affixed to the upper inner surface of the container, or home, of the male, who remains to guard the cluster until the eggs hatch. During the incubation period the male not only drives away intruders, but keeps water circulating around the eggs and cleans away debris which might smother them.

Onespot fringeheads feed upon a variety of organisms, especially crustaceans, which drift, crawl, or swim past their homes. On rare occasions they will dart short distances away from their shelter to snatch a particularly choice morsel. Fringehead remains have been found in the stomachs of bass and rockcod, and probably they are eaten by others.

Capture data. — Onespot fringeheads will take a baited hook, and an estimated 150 or more are caught by sportfishermen each year. They are pugnacious to the end, however, and snap viciously at the angler attempting to remove the hook. A few are netted each year incidental to seine or trawl fisheries, but most are captured alive by aquarium enthusiasts or are collected by scientific personnel using ichthyocides. They adjust rapidly to life in an aquarium and are extremely interesting to watch.

Other family members. — Two other fringeheads are known from Californian waters, but all can be told apart by checking their spots:

1. If there are two ocelli in the dorsal fin, it is a sarcastic fringehead.
2. If there is one ocellus, it is a onespot fringehead.
3. If there are no ocelli, it is a yellowfin fringehead.

Meaning of name. — *Neoclinus:* new *Clinus; uninotatus:* one mark, for the single ocellus.

Reef Finspot
Paraclinus integripinnis (Smith, 1880)

Distinguishing characters. The typical body shape, somewhat pointed snout, and long dorsal fin, comprised entirely of spines, which is elevated (saillike) anteriorly and has a jet black ocellus between spines 22 and 27, are sufficient to distinguish the reef finspot from all other fishes in our waters.

Natural history notes. – *Paraclinus integripinnis* has been captured in most rocky habitat between Santa Cruz Island and Serena Cove, Santa Barbara County, California, and Almejas Bay, Baja California (just south of Magdalena Bay). It is most abundant intertidally and in the shallow subtidal zone, but has been taken as deep as 50 feet (15.2 m). The maximum reported length for a reef finspot is 2-1/2 inches (6.4 cm).

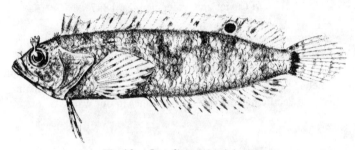

Fig. 18.   *Paraclinus integripinnis*

An examination of their otoliths indicates that most reef finspots will spawn when they are a year old. The oldest of several individuals about 2 inches (5.1 cm) long was three, so their maximum age probably does not exceed five. Spawning season and behavior are not known.

The few stomachs which have been examined have contained isopods and amphipods (crustaceans) exclusively. There is no record of a predator having fed upon reef finspots, but there is no reason to believe they would not be fair game for any carnivore in their environment.

Capture data. – Reef finspots are frequently captured alive by scuba divers, for use in marine aquaria. Most captures, however, have been made by biologists and museum and university personnel while making surveys of the biota for institutional collections. These little fishes adapt very well to life in an aquarium, and could afford an excellent opportunity to learn a great deal more about their ways than is now known.

Other family members. — For information on other clinids, please refer to the checklist and references in the appendices.

Meaning of name. — *Paraclinus:* near *Clinus; integripinnis:* entire fin — in allusion to the undivided dorsal fin comprised entirely of spines.

Clupeidae (Herring Family)
Pacific Herring
Clupea pallasii Valenciennes,
in Cuvier and Valenciennes, 1847

Distinguishing characters. — The sardinelike body, single dorsal fin, lack of striations on the gill cover, lack of a filamentous last dorsal ray, lack of sharp sawtoothed scales on breast and belly, and the pelvic fin positioned below the dorsal are sufficient to distinguish the Pacific herring from other fishes in our waters.

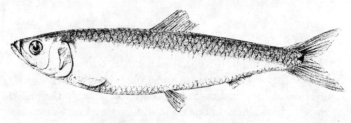

Fig. 19. *Clupea pallasii*

Natural history notes. — *Clupea pallasii* is a schooling species which ranges in the north Pacific from central Japan to the Bering Sea and San Diego, but at present it is not abundant south of San Francisco Bay. In former years, Pacific herring were found as far south as Ensenada, Baja California. Although reported to attain a length of 18 inches (45.7 cm), few are seen which exceed 15 inches (38.1 cm), and off California a 12-inch (30.5 cm) herring would be a giant. In the San Francisco and Tomales Bay areas, where most of the Californian populations are found, they attain a maximum age of about nine or ten years; in the Pacific northwest some herring are said to reach nineteen years. They mature when two, three, or four years old.

In our waters, schools of herring move into bays and estuaries during late winter, and spawn intertidally from then until early spring. Their adhesive eggs cling in great masses to eelgrass, kelp, and other fixed objects, where they are not only fed upon by great numbers of finned and feathered predators but are harvested (under permit) by commercial fishermen. The number of eggs per female herring depends upon her size (and age), but it ranges from an estimated 10,000 to an excess of 125,000. The relatively large eggs, averaging about 1/20-inch (1.3 mm) in diameter, hatch in about ten days.

Larval and juvenile herring have been found many dozens of miles at sea, presumably having drifted there with the currents. They are fed upon at all sizes by almost every type of predator there is, whether fish, bird, or mammal, and are much sought after by man for both food and bait. At all life stages, herring feed almost exclusively on planktonic organisms, which they are capable of straining from the water or selecting as desired.

Otoliths from *C. pallasii* have been found in many fossil deposits of Pliocene and Pleistocene age, and both otoliths and scales have turned up in a number of coastal Indian middens.

Capture data. — In California, the commercial catch has ranged from a peak of about 9 million pounds (4.1 million kg) in 1952 to less than 200 thousand (90,000 kg) in several different years. In Canadian waters, the annual herring harvest has reached nearly half a billion pounds (226 million kg) on occasion.

Other family members. — For information on the seven members of this family which are found in our waters, please refer to the checklist and references in the appendices and to our account of the sardine in *Marine Food and Game Fishes of California.*

Meaning of name. — *Clupea:* ancient name for herring; *pallasii:* in honor of Petrus Simon Pallas, an early naturalist and ichthyologist.

43

Cottidae (Sculpin Family)
Rosylip Sculpin
Ascelichthys rhodorus Jordan and Gilbert, 1880

Distinguishing characters. — This is the only cottid which lacks pelvic fins. It is dusky or olive-brown on the back and sides and lighter below; the margin of the dorsal fin is bright red, as are the inner edges of the thickened lips. There are no scales on the body.

Fig. 20. *Ascelichthys rhodorus*

Natural history notes. — *Ascelichthys rhodorus* ranges from Sitka, Alaska, to Pillar Point, San Mateo County, California. They inhabit the rocky intertidal and subtidal zones almost exclusively, and in some areas are extremely common. The maximum reported length is 5.9 inches (14.9 cm), but we have no information on maximum age. A specimen slightly shorter than 4 inches (10.2 cm) long was three years old, while one an inch (2.5 cm) longer was five.

None of the stomachs we examined contained food, and we have no information on reproduction or reproductive behavior. Remains of a rosylip sculpin were found in the stomach of a rockcod, but we have no information on other predators.

Capture data. — Rosylip sculpins can be caught alive quite easily, and fair numbers are thus taken for use in aquaria. Mostly, however, they are collected by scientific personnel who are studying the biota of rocky intertidal and subtidal areas, or who are making collections for museums or universities.

Other family members. — For information on the more than forty sculpins which inhabit our marine waters, please refer to the checklist and references in the appendices and to our account of the cabezon in *Marine Food and Game Fishes of California.*

Meaning of name. — *Ascelichthys:* literally, a fish without legs — pelvic fins; *rhodorus:* rose margin, referring to the color of the lips.

Woolly Sculpin
Clinocottus analis (Girard, 1858)

Distinguishing characters. — A combination of characters which will separate the woolly sculpin from all other marine fishes in our waters includes the following: typical cottid body shape; pelvic fins comprised of 1 spine and 3 soft rays; 8 to 10 spines and 15 to 18 rays in the dorsal fin; gill membranes which are broadly united and free from the isthmus; cirri and minute scales between the dorsal fin and lateral line; anus located midway between the pelvic fins and anal-fin origin; and a bifid or trifid preopercular spine.

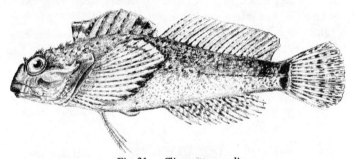

Fig. 21.   *Clinocottus analis*

Natural history notes. — *Clinocottus analis* ranges from near Cape Mendocino, California, to Asuncion Point, Baja California, and offshore at Guadalupe Island. It has been reported from depths of 60 feet (18.3 m) but is most abundant in the intertidal and shallow subtidal zones. Most individuals are a dusky green or grayish color but some are reddish, depending upon dominant habitat colors. They have been reported to attain a maximum length of 7 inches (17.8 cm), but few are seen which exceed 5 (12.7 cm).

Examination of their otoliths indicates that at lengths of 4 inches (10.2 cm) they are two years old, and at 5 inches (12.7 cm) they are three. We have no information on maximum age, or age at first maturity, nor do we have information on spawning season, spawning behavior, or other facets of reproduction, except for a note that a cluster of several hundred of their eggs found under rocks required 24 days to hatch.

The few stomachs which have been examined have contained mostly algae, crustaceans, and polychaete worms, plus a few tiny marine snails.

Capture data. — Woolly sculpins are abundant in rocky tidepools, and with a minimum of effort they can be caught by hand or with small handheld dipnets. They are quite hardy and adapt well to an aquarium. Most records, however, are based upon specimens collected by biologists or museum and university personnel using ichthyocides in the rocky intertidal zone.

Other family members. — For information on the more than forty sculpins which inhabit our marine waters, please refer to the checklist and references in the appendices.

Meaning of name. — *Clinocottus: Clinus + Cottus,* two genera of shallow-water fishes, which are similar to some sculpins; *analis:* pertaining to the large anal papilla of the male.

Pacific Staghorn Sculpin
Leptocottus armatus Girard, 1854

Distinguishing characters. — The typical cottid shape, broad and somewhat flattened head, relatively large mouth, deeply

Fig. 22.   *Leptocottus armatus*

46

notched dorsal fin, antlerlike preopercular spine, lack of scales on body, and general color are more than sufficient to distinguish a staghorn sculpin. It is typically greenish-gray above, shading to white below, with yellowish coloring around the mouth and along the sides; there also is a series of dusky bars on the pectoral fins. They are very slimy to handle.

Natural history notes. — *Leptocottus armatus* ranges from Chignik, Alaska, to San Quintin Bay, Baja California. Although reported to attain a length of 12 inches (30.5 cm) in our waters, and 18 inches (45.7 cm) in Canadian, the largest for which we can find an actual measurement was just under 10 inches (24.8 cm) and weighed about one-half pound (220 g). Staghorn sculpins have been reported from depths as great as 510 feet (155.5 m) but few are seen deeper than about 60 (18.3 m). They are most abundant in shallow subtidal areas of the outer coast, and in bays and estuaries. They sometimes ascend freshwater streams for short distances.

They become mature when one year old, and although maximum age is unknown, individuals 8 to 9 inches (20.3 to 22.9 cm) long were three years old, and one slightly less than 10 inches (25.4 cm) was five. Spawning takes place primarily between October and April, and an average-size female will lay about 5,000 eggs in a season. The eggs hatch in about ten days, depending upon water temperatures. The most abundant food items found in their stomachs are crustaceans (crabs, shrimp, and amphipods), but they also eat many kinds of larval, juvenile, and adult fishes, as well as polychaete worms, mollusks, and other invertebrates. In turn, staghorn sculpins are eaten by a variety of birds, marine mammals, and fishes.

Their remains have been found in fossil deposits of Pliocene and Pleistocene age, and in several coastal Indian middens.

Capture data. — Staghorn sculpins are caught in relatively low numbers by commercial fishermen, and only incidentally to other fisheries. Sportfishermen, however, catch an estimated 20,000 per year, primarily while fishing from piers. They are such an undesirable fish in the sportfisherman's bag that we doubt if many are used for food. During a two-year trawling study of the Sacramento-San Joaquin River estuary, the 2,644 staghorn sculpins which were caught in San Pablo and Suisun bays made them the most abundant of the bottom-dwelling fishes taken. They adjust well to life in an aquarium, but cannot be put in with smaller fishes of other species.

Other family members. – For information on the more than forty sculpins which inhabit our marine waters, please refer to the checklist and references in the appendices.

Meaning of name. – *Leptocottus:* slender *Cottus; armatus:* armed, in reference to the sharp preopercular spines.

Sailfin Sculpin
Nautichthys oculofasciatus (Girard, 1857)

Distinguishing characters. – This is the only fish in our marine waters with a sail-like first dorsal fin where several of the rays are of about equal length, and with a velvety skin texture. The velvety skin is caused by the tiny, highly modified ctenoid scales. The body color is yellowish-brown or yellowish-gray, and there is a dark brown stripe running through the eye.

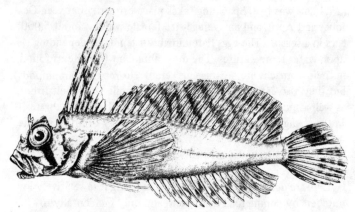

Fig. 23. *Nautichthys oculofasciatus*

Natural history notes. – *Nautichthys oculofasciatus* has been recorded from Anadyr Gulf on the east coast of the Bering Sea to Point Sal and San Miguel Island, California. They usually are found in moderately deep water but have been taken from the intertidal zone up to 413 feet (125.9 m) of water. They inhabit rocky areas where the algal coverage is dense, and are believed to be primarily nocturnal in their habits. Maximum recorded size is variously reported as 6.8 and 8 inches (17.3 and 20.3 cm), depending upon the authority. Since a fish which was 6.8 inches long (17.3 cm) and weighed about 2-1/2 ounces (70 g) was only two years old, we prefer to believe that 8 inches (20.3 cm) would be closer to maximum size. Several individuals ranging in length from 5-1/4 to 6-1/4 inches (13.3 to 15.8 cm) were a year old, judged by growth rings on their otoliths, and were mature. Spawning takes place in late winter and early spring, and the orange-colored egg mass adheres to the substratum.

The few stomachs which have been examined have contained crab remains, for the most part, but other crustaceans and some fish remains also were found. In turn, sailfin sculpins are fed upon by several predatory fishes which live in the same habitat.

Capture data – Sailfin sculpins have been caught on hook and line, but only rarely. They are not too difficult for scuba divers to catch alive with small handheld dipnets, but most records for the species are based upon specimens which were collected by biologists or museum and university personnel using ichthyocides. They are a hardy fish and adapt well to life in an aquarium.

Other family members. – For information on the more than forty sculpins which inhabit our marine waters, please refer to the checklist and references in the appendices.

Meaning of name. – *Nautichthys:* sailor fish, in allusion to the sail-like first dorsal fin; *oculofasciatus:* eye-banded.

Grunt Sculpin
Rhamphocottus richardsonii Günther, 1874

Distinguishing characters: – The peculiar shape of this odd little fish will distinguish it from all others inhabiting our marine waters. In addition, its skin is velvety to the touch, because of highly modified scales. The body color is yellowish with irregular brownish streaks, and the fins are reddish-orange. They swim awkwardly, with the head up, but mostly they are observed "crawling" along the bottom in a series of short jumps. They make grunting noises when handled.

Natural history notes. – *Rhamphocottus richardsonii* ranges from the Bering Sea to Santa Monica Bay and 18 miles (28.9 km) southeast of San Nicolas Island. They are most abundant in the intertidal and shallow subtidal zones in the northern part of their range, but have been caught as deep as 600 feet (182.8 m) in southern areas. They attain a maximum length of 3-3/8 inches (8.5 cm). A grunt sculpin just over 2-1/2 inches long (6.8 cm) was two years old, judged by rings on its otoliths, while one that was 3-3/8 inches (8.5 cm) appeared to be five.

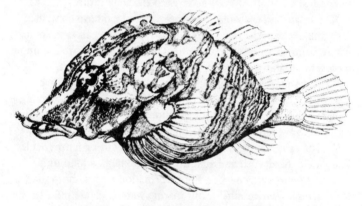

Fig. 24.   *Rhamphocottus richardsonii*

Spawning takes place from about August through October, and it is the female which is the aggressor. In an aquarium, the female pursues the male until she can trap him in a niche, small rocky cavern, or similar cul-de-sac. She keeps him trapped until she lays her yellow or orange-colored eggs (about 150 of them) and he fertilizes them. The eggs require sixteen to twenty weeks to hatch, depending upon water temperatures.

Grunt sculpins eat mostly crustaceans, but small fishes and polychaete worms have also been found in their stomachs. In turn, grunt sculpins are preyed upon by several kinds of rockcod and other voracious fishes which inhabit the same environment.

Capture data. – Grunt sculpins can and have been caught on hook and line, and a few individuals also have been taken in fine-mesh trawl nets, incidental to other fisheries. Mostly, however, they are taken in the course of scientific investigations by personnel using ichthyocides, traps, and other gear. They are easy to capture alive using scuba gear and handheld nets, and they add comedy to any saltwater aquarium. If the water temperature in the aquarium is maintained below 55°F (12.8°C), grunt sculpins will survive and provide comical viewing for several years.

Other family members. – The grunt sculpin in past years sometimes has been listed in its own family (Rhamphocottidae), but the recent trend is to consider it a member of family Cottidae. For information on the more than forty sculpins which inhabit our marine waters, please refer to the checklist and references in the appendices.

Meaning of name. – *Rhamphocottus:* snout *Cottus,* or a *Cottus-*like fish with a snout; *richardsonii:* for John Richardson, early naturalist and explorer.

Cynoglossidae (Tonguefish Family)
California Tonguefish
Symphurus atricauda (Jordan and Gilbert, 1880)

Distinguishing characters. – This is the only flatfish in our waters which lacks a distinct caudal fin; the dorsal and anal fins are contiguous with the caudal, and taper to a point. In addition,

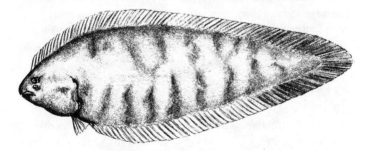

Fig. 25. *Symphurus atricauda*

the eyes are on the left side, the mouth is weak and curved downward, and the projecting strongly ctenoid scales are so arranged that it is impossible to hold a live wriggling tonguefish with one hand.

Natural history notes. – *Symphurus atricauda* ranges from Big Lagoon, Humboldt County, California, to Cape San Lucas, Baja California. They inhabit sandy mud or mud bottoms and have been captured in depths between 5 and 660 feet (1.5 to 201 m). Maximum reported length is 8 1/4 inches (20.9 cm); this fish weighed just over 2 ounces (67 g), but we have no information on its age.

Spawning occurs primarily during June through September, but we have no information on the time required for the eggs to hatch, number of eggs per female, age at maturity, maximum age, etc.

Several stomachs which we examined contained the remains of crustaceans and polychaete worms, but many other kinds of invertebrates doubtless are eaten also. Tonguefish remains have been found in a variety of other fish stomachs, particularly other flatfishes, and probably are fed upon by some birds such as cormorants.

Fossil otoliths from *S. atricauda* have been found in several Pliocene and Pleistocene deposits in California, and there is a skeletal imprint of one from Miocene shales of Palos Verdes.

Capture data. – Tonguefishes have been reported in the sport-fisherman's bag, but only on rare occasions when very tiny hooks were used. Small numbers are caught accidentally or incidentally in commercial trawl fisheries, but the mesh used in most nets is too large to take them regularly. They were the third most abundantly caught fish during a six-year study in Santa Monica Bay, where fine-mesh trawls were used. They comprised 10,438 of nearly 113,000 fishes netted during this study.

Other family members. – *S. atricuada* is the only member of the family known off California, or within several hundred miles (500-700 km).

Meaning of name. – *Symphurus:* a combination of three words – together, grow, and tail – alluding to the tail fin being a continuation of the dorsal and anal fins; *atricauda:* black tail.

52

Diodontidae (Porcupinefish Family)
Pacific Burrfish
Chilomycterus affinis Günther, 1870

Distinguishing characters. — This is the only porcupinefish in which the projecting spines are immovable. When endangered or harassed it is able to swallow water or air and inflate itself like a balloon. It is dusky or bluish-black above, white on the underside. Its parrotlike "beak" is the result of fused jaw teeth.

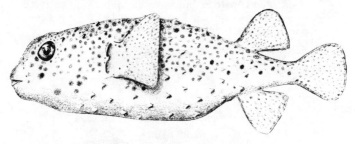

Fig. 26. *Chilomycterus affinis*

Natural history notes. — *Chilomycterus affinis* ranges from San Pedro to Peru and the Galapagos Islands, but no more than three or four individuals have been seen or captured in our waters. They typically inhabit shallow rocky or coral reef areas in tropical and subtropical waters. They are reported to reach a length of about 20 inches (50.8 cm), but we do not have any length or weight information for large individuals. A 10-1/2-inch (26.7 cm) fish weighed 1-1/2 pounds (680 g).

We know that they brace themselves against the substrate to sleep at night, and will eat bits of fish or squid in an aquarium, but we have no information concerning their reproductive behavior, food habits, age, and other facets of life history. They are preyed upon, especially at small size, by such large predators as wahoos, tunas, serranids, and sharks.

Capture data. — They are easily captured by scuba divers, especially at night when they are asleep, in waters to the south of California. They are taken incidentally with nets such as beach seines, gill nets, purse seines, and lamparas. Because they are rather feeble swimmers, many are cast ashore by storm waves. They are used commercially as curios by inflating and drying, but they make interesting aquarium inhabitants, especially when small.

Other family members. – One other member of the family has been captured or found in our waters on two or three occasions. The two species are easily distinguished:

1. If the spines on the body are short, sparsely distributed, immovable, and typically have three basal roots, it is a Pacific burrfish.
2. If the spines are long, numerous, movable (can be raised and lowered), and typically have two basal roots, it is a spotted porcupinefish.

Meaning of name. – *Chilomycterus:* lip nose; *affinis:* related – to associated species.

Embiotocidae (Surfperch Family)
Shiner Perch
Cymatogaster aggregata Gibbons, 1854

Distinguishing characters. – The typical perchlike shape, concave profile above the eye, and body color will usually distinguish this little fish at a glance. There are three yellow bars on each side, interspersed with blackish or dusky areas. Pier, jetty, and dock fishermen call them "seven-eleven perch," because of the resemblance these bars have to the number 711.

Natural history notes. – *Cymatogaster aggregata* ranges from Port Wrangell, Alaska, to San Quintin Bay, Baja California. They typically form loose schools or aggregations in shallow nearshore waters along the outer coast, and also abound in bays and estuaries sometimes entering freshwater. Maximum reported size is 7 inches (17.8 cm), and maximum reported depth of occurrence is 480 feet (146.3 m).

Males apparently are sexually mature at birth, and females at one year of age. Mating takes place during the entire year, but young are born primarily during spring and summer months. A pregnant female 6-3/4 inches long (16.9 cm) which weighed just under 3 ounces (75 g) contained 20 term young averaging about

54

1-1/2 inches long (4.1 to 4.3 cm) and weighing a total of just less than one ounce (22 g). Smaller females have progressively fewer young. The oldest of 327 individuals for which ages were determined was three years; maximum age probably does not exceed four.

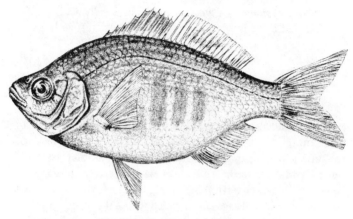

Fig. 27.   *Cymatogaster aggregata*

Shiner perch are omnivorous in their eating habits. Their stomachs have been found to contain primarily crustaceans and algae, but annelid worms and mollusk parts also occur rather regularly. Birds, fishes, marine mammals, and crabs feed heavily on shiner perch at times.

Shiner perch otoliths have been found in many West Coast fossil deposits of Pliocene and Pleistocene age, and have been present in several Indian middens.

Capture data. – *C. aggregata* is too small to be of commercial value for human consumption, so the incidental catch is primarily used for food by sportfishermen, who catch an estimated 325,000 per year in our waters, mostly from piers, docks, breakwaters, and jetties. Each year countless thousands are drawn into electric steam generating plants with cooling waters and are destroyed, because there is no way for fishes to get out. They are small enough and hardy enough to make an interesting addition to a saltwater aquarium.

Other family members. – For information on the eighteen other embiotocid perches which inhabit our marine waters, please refer to the checklist and references in the appendices and to our accounts in *Marine Food and Game Fishes of California.* One embiotocid perch which is native to California lives entirely in fresh water.

Meaning of name. – *Cymatogaster:* foetus-belly – alluding to the fact that it is a live-bearer; *aggregata:* crowded together.

Black Perch
Embiotoca jacksoni Agassiz, 1853

Distinguishing characters. – The perchlike body shape, enlarged scales on the side between pectoral and pelvic fins, and coloration are sufficient to distinguish black perch from all other fishes in our waters. When fresh, the body is blackish-brown to orangish-brown, with darker vertical bars along the sides; the fleshy lips are a deep yellow-orange (from which it gets the name buttermouth); and the anal fin has alternating blue and yellow-orange stripes which are most intense during breeding season.

Natural history notes. – *Embiotoca jacksoni* ranges from Fort

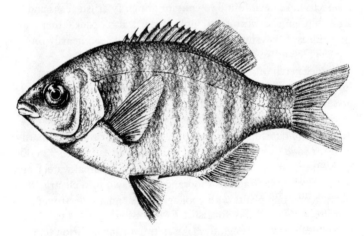

Fig. 28. *Embiotoca jacksoni*

Bragg, California, to Abreojos Point, Baja California, and is also found offshore at Gudalupe Island. They occur in loose schools or aggregations in both sandy and rocky habitat along the outer coast and in bays and estuaries. Black perch usually are found at depths shallower than 80 feet (24.4 m), but have been captured as deep as 150 feet (45.7 m). Maximum length is 15-1/3 inches (38.8 cm); a 13-inch-long (33.0 cm) male weighed about 1-1/2 pounds (658 g).

Males and females are mature at one year of age and slightly under 5 inches (12.7 cm) in length. Some mating takes place during all months of the year, but most young are born during spring and summer months. A number of four-year-old females averaging about 9 inches (22.8 cm) long contained 5 to 26 young, averaging 13. As with other members of the family, the larger (older) the female, the greater the number of young per pregnancy. Birth size averages about 2 inches (4.5 to 5.7 cm). Black perch apparently live for ten years or longer; a 12-1/2-inch-long (31.8 cm) male was nine years old, and several larger individuals which were not measured were ten.

Numerous stomachs have contained crustaceans, algae, mollusks (mostly mussels and clams), bryozoans, and polychaete worms. Scuba divers often have seen black perch eating parasites, which they were removing from other kinds of fishes. Remains of large black perch have been found in stomachs of electric rays, and juveniles are fed upon by a variety of voracious fishes and by birds such as cormorants.

Black perch otoliths have been found in several Pliocene and Pleistocene deposits and in coastal Indian middens.

Capture data. – Black perch make up an unknown portion of the commercial perch catch in California, which is primarily taken with gill nets and lamparas. An estimated 133,000 are caught by sportfishermen each year, including skindivers. Both the commercial and sport catch are utilized for human consumption. Small black perch make interesting additions to home aquaria, but adults are too large for most private aquaria.

Other family members. – For information on the eighteen other embiotocid perches which inhabit our marine waters, please refer to the checklist and references in the appendices.

Meaning of name. – *Embiotoca:* living and bringing forth – alluding to their bringing forth living young; *jacksoni:* for A. C. Jackson of San Francisco, who first noted that black perch bore living young and brought this information to the attention of Alexander Agassiz.

Dwarf Perch
Micrometrus minimus (Gibbons, 1854)

Distinguishing characters. – The perchlike shape, body coloration, 12 to 16 soft dorsal rays, and body depth relative to length will distinguish this small embiotocid from other Californian marine fishes. When freshly caught, it is greenish-blue or steely-blue above and silvery on the belly. There is irregular dusky mottling and blotching on the sides and an interrupted longitudinal dark band along each side. The axil of the pectoral fin is very black.

Natural history notes. – *Micrometrus minimus* ranges from Bodega Bay, California, to Cedros Island, Baja California. They occur in small loose aggregations or schools in the intertidal and

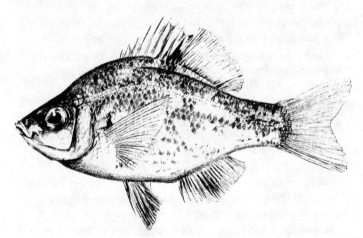

Fig. 29. *Micrometrus minimus*

shallow subtidal zones into depths of 30 feet (9.1 m). Dwarf perch seldom stray too far from rocky habitat, including break-waters and jetties, where they mostly are found in and around surfgrass beds. Maximum size is 6-1/4 inches (16.1 cm) and about 2-1/2 ounces (76 g); this fish, a female, was three years old.

Males are mature at birth, but females do not mature until they are a year old and about 3-1/2 inches (8.9 cm) long. Young are born during late spring and early summer and are about 1 inch (2.5 cm) long. Of 63 individuals examined for age, no male exceeded one year, but many females were two, and two females were three.

Dwarf perch stomachs which we have examined have contained crustaceans (especially amphipods), mollusks, polychaete worms, and much algae. We do not know of any specific predator on *M. minimus,* but suspect they would be fair game for most large predators living in their environment.

Capture data. – A few dwarf perch are caught by sportfishermen each year, but most specimens on record were taken by scientific personnel who were making general fish collections or studying the biota of rocky intertidal and subtidal areas. Dwarf perch are quite hardy and small enough to make colorful and interesting additions to a marine aquarium.

Other family members. – For information on the eighteen other embiotocid perches which inhabit our marine waters, please refer to the checklist and references in the appendices.

Meaning of name. – *Micrometrus:* small mother; *minimus:* smallest.

Engraulidae (Anchovy Family)
Deepbody Anchovy
Anchoa compressa (Girard, 1858)

Distinguishing characters. – The anchovylike shape, including overhanging "nose," semitransparent body with a broad silver stripe along the midside, and 29 to 33 anal rays will distinguish the deepbody anchovy from all other marine fishes in our waters.

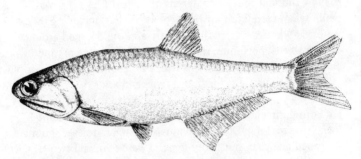

Fig. 30. *Anchoa compressa*

Natural history notes. – *Anchoa compressa* ranges from Morro Bay, California, to Todos Santos Bay, Baja California. Though a schooling species, they are never found in dense aggregations. Deepbody anchovies primarily are found in the farthest reaches of back bays, lagoons, and estuaries, but they do occur in limited numbers along the outer coast where the water is not overly turbulent. Maximum size appears to be a 6-1/2-inch-long (16.5 cm) female, which weighed just over 1 ounce (31 g) and was seven years old.

Deepbody anchovies contain nearly ripe eggs during most months of the year, but successful spawning may be confined mostly to spring and early summer. There is no information on other facets of reproduction, including number of eggs for various ages (sizes), time required to hatch, etc.

Deepbody anchovies are basically filter feeders, but at times have been observed feeding by selection (i.e., they do not just take what they can filter from the water, but actually pick out certain microorganisms which appeal to them).

Otoliths of *A. compressa* have been found in several Pleistocene deposits in southern California.

Capture data. – We know of one instance where a deepbody anchovy was caught with a tiny hook baited with angleworm, but this must be considered unusual. Mostly they are not seen except by fishery biologists and other scientific personnel who are collecting backbay fishes with fine-mesh seines and other effective gear.

Other family members. – Although four other anchovies have

60

been reported from our waters, one of these (the slim anchovy) has been reported only once, and one other (the anchoveta) is rare this far north.

1. If the gill covers are joined to each other in the throat region, it is an anchoveta.
2. If the body is nearly round in cross-section, the head is longer than the body is deep, and the color is a dusky blue or green above, it is a northern anchovy.
3. If the body is laterally compressed, the head is shorter than the body is deep, the color is whitish above, and
   a. there are 17 to 20 anal rays, it is a slim anchovy.
   b. there are 23 to 26 anal rays, it is a slough anchovy.
   c. there are 29 to 33 anal rays, it is a deepbody anchovy.

Meaning of name. — *Anchoa:* ancient name for anchovy; *compressa:* compressed.

Ephippidae (Spadefish Family)
Pacific Spadefish
Chaetodipterus zonatus (Girard, 1858)

Distinguishing characters. — The unique body and fin shape and grayish or bluish-gray background color, with overlying broad dark bars, are sufficient to distinguish the Pacific spadefish from all other fishes in our waters. In very large adults, however, the dark side bars are indistinct and very difficult to see.

Natural history notes. — *Chaetodipterus zonatus* has been recorded from San Diego to northern Peru, but it is not abundant north of about Magdalena Bay, and the Californian records are more than 100 years old. The maximum reported size is "to about two feet" (60.9 cm), or "to about 25 inches" (62.5 cm), but these reports cannot be referred to an individual fish. The largest specimens we have seen were caught in gill nets in water 90 to 150 feet deep (27.4 to 45.7 m), and although far short of the reported "two feet" they exceed by more than 4 inches (10.2 cm) the largest for which we can find an authentic record. The largest of these gillnet-caught spadefishes, a female without dark bars, was 12 inches long (30.5 cm) and weighed just under

61

1-3/4 pounds (775 g). This fish was more than nine years old judged by growth zones on its otoliths.

We have no information on age at first maturity, spawning season, fecundity, spawning behavior, time needed for the eggs to hatch, etc. Many stomachs we have examined were empty, but others have contained an assortment of food items. Those from deep water have contained only pelagic red crabs (*Pleuroncodes planipes*), but those from shallow water have contained sponges, polychaete worms, gorgonians, tunicates, anemones, mollusks, and algae.

Large swollen bony elements found in some Miocene deposits which have been attributed to *Chaetodipterus* are probably not from *C. zonatus*.

Fig. 31.   *Chaetodipterus zonatus*

Capture data. – Where most abundant, Pacific spadefish can be and frequently are caught on hook and line. They are also netted and since the flesh is edible, they occasionally can be found in Mexican and Central American fish markets. They make excellent additions to aquaria, but are a bit large for the average home aquarium.

Other family members. – No other member of the family is known within many hundreds of miles of California.

Meaning of name. – *Chaetodipterus: Chaetodon* two-fin – refering to the *Chaetodon*-like body with a divided dorsal fin; *zonatus:* zoned or banded.

Gerreidae (Mojarra Family)
Pacific Flagfin Mojarra
Eucinostomus gracilis (Gill, 1862)

Distinguishing characters. – The perch-shaped body, overall silvery color, protractile mouth, and tricolored first dorsal fin are sufficient to distinguish this small fish. The first dorsal fin is black at both tip and base and has a milky white central area.

Natural history notes. – *Eucinostomus gracilis* has been recorded from Anaheim Bay, California, to Callao, Peru, but is not abundant

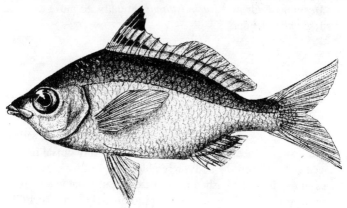

Fig. 32. *Eucinostomus gracilis*

north of Magdalena Bay, Baja California. In fact, it has been taken in our waters only once, in February 1965. Flagfin mojarras typically inhabit shallow inshore areas both on the open coast and in bays and estuaries where the bottom is sandy or muddy. They travel in loose schools or aggregations containing from 10 to 50 or more individuals. Maximum reported size is 8.2 inches (20.8 cm); the Anaheim Bay fish was 7 inches long (17.8 cm).

We have no information on ages or reproduction, and very little on food. Several stomachs which we examined contained an assortment of benthic invertebrates, primarily polychaete worms, mollusks, crustaceans, and bryozoans.

Capture data. — Where they are abundant, these small mojarras are caught mostly with beach seines, which are set and pulled from shore. The one individual caught in our waters was taken on a small hook baited with clam. The flesh is edible, but their generally small size makes them undesirable as table fare. They are interesting to watch and readily become acclimated to aquarium life.

Other family members. — One other mojarra has been taken in our waters on one occasion. The two are easily distinguished:

1. If the first dorsal fin is plain-colored (slightly dusky), it is a silver mojarra.
2. If the first dorsal fin is tricolored (black at tip and base, and milky white centrally), it is a flagfin mojarra.

Meaning of name. — *Eucinostomus:* literally, to move the mouth well; *gracilis:* slender.

Gobiesocidae (Clingfish Family)
Slender Clingfish
Rimicola eigenmanni (Gilbert, 1890)

Distinguishing characters. — This tiny kelp-colored fish with a suction pad in place of pelvic fins, and a single tiny dorsal fin placed well back on the body, resembles only its own close relatives. It can be distinguished from these, because the two pairs of pores near the tip of the snout are approximately in a straight line, and there are 17 to 19 pectoral rays.

Fig. 33.   *Rimicola eigenmanni*

Natural history notes. – *Rimicola eigenmanni* ranges from Palos Verdes Peninsula, California, to San Juanico Bay, Baja California. It associates primarily with surfgrass and various shallow-water kelps, and clings to these with its modified suction-pad pelvic fins. They are perfectly camouflaged with their surroundings, and since they rarely move unless harassed, they are extremely difficult to find in their natural surroundings. Maximum reported size is 2-1/4 inches (5.7 cm), and maximum recorded depth is 48 feet (14.6 m), but they are rare at depths exceeding 10 to 12 feet (3.0 to 3.7 m).

They presumably spawn during most of the year, judged by examination of ovaries of preserved females, and their rather large kelp-colored eggs adhere to the stems and blades of surfgrass and algae. The stomachs we have examined have contained tiny crustaceans almost exclusively. Slender clingfish apparently serve as food in a small way to larger predators in their own environment.

We have no information on age, but it has been speculated that slender clingfish rarely live more than a year.

Capture data. – Few slender clingfish are collected or seen by other than scientific personnel using ichthyocides in shallow reef areas. They probably would survive quite well in an aquarium if fed brine shrimp, but their sedentary habits would not make them very interesting to observe.

Other family members. – In all, seven species of clingfish are known from Californian waters. These seven species belong to only two genera: *Gobiesox* (four species), having a dorsal fin which reaches almost to the caudal fin and contains 10 to 16 rays; and *Rimicola* (three species), in which the dorsal fin is well ahead of the caudal and contains only 5 to 8 rays. To identify clingfish species requires a microscope and careful comparison of numerous characters, so we will not attempt to differentiate them here.

65

Meaning of name. – *Rimicola:* literally, a crevice inhabitant; *eigenmanni:* for Carl H. Eigenmann, renowned American ichthyologist.

Gobiidae (Goby Family)
Blackeye Goby
Coryphopterus nicholsii (Bean, 1882)

Distinguishing characters. – The blackeye goby is easily recognized by a combination of characters including the fused pelvic fins, large scales, a fleshy ridge on top of the head which extends from between the eyes to the dorsal fin origin, large dark eyes, and a black-edged first dorsal fin. The overall body color is a light tan with some brownish or greenish speckling.

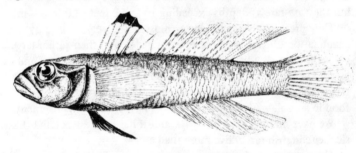

Fig. 34. *Coryphopterus nicholsii*

Natural history notes. – *Coryphopterus nicholsii* ranges from Skidegate Inlet, Queen Charlotte Islands, British Columbia, to Rompiente Point, Baja California. Larvae and postlarvae have been taken with plankton nets over 65 miles (104.6 km) offshore, where depths exceed several thousand feet (several hundred meters), but adults apparently have never been noted deeper than 348 feet (106 m). Maximum size is given as "about 6 inches" (15.2 cm) (California) and "4.75 inches" (12.1 cm) (Canada); the largest for which we have found a record was about 4-3/4 inches (11.7 cm), but we do not have a weight for this fish.

The male puts on a courtship display for the female of his

choosing and lures her into a nesting cave, which he has prepared under a suitable rock or ledge. Spawning takes place from April to October, and the eggs (about 1700 per nest) are stuck to the roof of the cave one layer thick. They are a "faded pink" color when first spawned, but darken as they develop. The male remains on guard during the period the eggs are incubating.

Blackeye gobies apparently mature after their first winter, but we have no information on maximum age. The oldest fish we have information on was just over 4-1/4 inches long (10.9 cm) and weighed a shade more than one-half ounce (16 g); it was four years old.

The stomachs we have examined have contained small gastropod mollusks and crustaceans almost exclusively. Remains of blackeye gobies have been found in albacore stomachs (postlarvae) and in bass and rockcod stomachs. They doubtless serve as forage for many other predators which inhabit their type of rocky environment.

Otoliths of *C. nicholsii* are abundant in several Pliocene and Pleistocene deposits in California.

Capture data. – Blackeye gobies can be and have been caught with tiny shrimp-baited hooks, but this is a slow and uncertain technique at best. Fishery biologists and other scientific personnel using ichthyocides in subtidal rocky environments are able to collect quantities of them, and most of our information is based upon such collections. They often are caught alive by scuba divers using slurp guns and various handheld nets, and they adapt well to life in a home aquarium.

Other family members. – Three of the thirteen kinds of gobies which inhabit our marine waters are covered in this book; for information on the other ten, please refer to the checklist and references in the appendices. A freshwater goby native to Japan has become established in the San Francisco Bay area and occasionally is taken in the Bay proper, but we have not included it in this book.

Meaning of name. – *Coryphopterus:* alluding to the fleshy crest which runs along the center line of the head to the dorsal fin; *nicholsii:* for Capt. H. E. Nichols, its discoverer.

67

Longjaw Mudsucker
Gillichthys mirabilis Cooper, 1864

Distinguishing characters. – The typical body shape, in conjunction with the completely united pelvic fins, easily distinguish gobies. Among several look-alike gobies, the longjaw mudsucker attains a much larger size as an adult, and its "trademark" is the very large mouth with a greatly developed maxillary which often reaches the margin of the gill opening. Longjaw mudsuckers are olive-brown above and lighter below; the sides, back, and fins are speckled or mottled with darker shades.

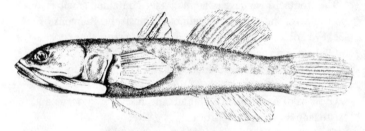

Fig. 35.   *Gillichthys mirabilis*

Natural history notes. – *Gillichthys mirabilis* ranges from Tomales Bay, California, to about Magdalena Bay, Baja California, and is also found in the central and upper Gulf of California and the Salton Sea. It typically inhabits very shallow mudflat areas of bays and lagoons, but has been taken in 35 feet (10.7 m) of water in the Salton Sea. The largest known individual was 8.2 inches (20.8 cm) long and weighed 6.2 ounces (175 g).

After a courtship display by the male, which may last a day or longer, the female deposits her eggs in a cleaned-out depression or nest prepared by the male. Spawning takes place from January through September, at least, and the eggs require ten to twelve days to hatch, depending upon water temperature. Depending upon her size, a single female will extrude between 4,000 and 9,000 eggs per spawning, and usually spawns two or three times per season at forty- to fifty-day intervals. The newly hatched larvae lead a short pelagic life before settling down in suitable habitat.

Females which hatch in January or February will contain mature eggs in August or September of the same year. They grow to about 5 inches (12.7 cm) their first year, and add an inch or two (2.5 to 5.1 cm) their second year. Few longjaw mudsuckers live longer than three years.

Polychaete worms, tiny mollusks, and other worms have been the most important items noted in their stomachs. Mudsuckers provide forage for several kinds of shorebirds, and for a few predaceous fishes, including an occasional stingray, which frequent intertidal mudflat areas.

Otoliths of *G. mirabilis* have been found in Pleistocene deposits near San Diego.

Capture data. – A relatively few mudsuckers are collected by scientific personnel during studies of backbay and estuarial environments. A flourishing commercial fishery also exists, however, and as many as 14,000 pounds are taken and sold as live bait each year. Mudsuckers harvested by this live-bait fishery are taken almost exclusively in small cylindrical wire traps with a fyke in one end. As long as they are kept cool and moist, mudsuckers will live for several days out of water.

Other family members. – Three of the thirteen kinds of gobies which inhabit our marine waters are covered in this book; for information on the other ten, please refer to the checklist and references in the appendices.

Meaning of name. – *Gillichthys:* Gill fish – literally, a fish named in honor of Theodore Gill, a renowned early American ichthyologist; *mirabilis:* wonderful.

Blind Goby
Typhlogobius californiensis Steindachner, 1880

Distinguishing characters. – This pink-colored goby has the characteristic fused pelvic fins, but lacks both scales and eyes. They live in pairs in shrimp burrows, and can be recognized for this symbiotic behavior as well as for physical features. The eyes degenerate in adults and become difficult to see, hence the description "blind."

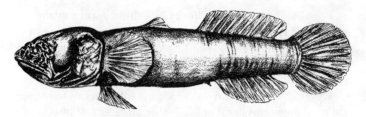

Fig. 36. *Typhlogobius californiensis*

Natural history notes. – *Typhlogobius californiensis* ranges from San Simeon Point, California, to Magdalena Bay. They inhabit the intertidal and shallow subtidal zones to a depth of 25 feet (7.6 m) or more. Maximum reported size is 3-1/4 inches (8.3 cm).

They pair off before they are six months old and move into the burrows of a ghost shrimp, *Callianassa affinis,* where they live out their lives. Even though his mate is living with him, the male goes through a courtship display prior to spawning. Eggs are laid from May through July, and the "nest" is a cleaned solid surface (rock or shell) within the shrimp burrow. Depending upon her size, the female attaches between 1500 and 2500 eggs to the nest substrate, and both parents keep the eggs cleaned of debris and guard the nest until hatching takes place, ten to twelve days after fertilization. Each female will spawn from two to four times per season. The newly hatched larvae have eyes and drift out of the shrimp burrow for a short pelagic life before settling down to raise their own families. It is believed that maximum age for a blind goby is three years.

They feed upon detritus and debris which drifts or is swept into the *Callianassa* burrow, and apparently obtain most nourishment from bits of animal flesh and seaweed comprising part of the debris.

Capture data. – Many blind gobies are dug up accidentally by persons digging bait (worms or shrimp) in a cobble or rocky habitat along the outer coast. They are sluggish and can tolerate very stagnant water for long periods of time, so many are collected for home aquaria. Occasionally scientific personnel will collect them for ongoing studies, or in enumerating the biota of an intertidal area.

Other family members. – Three of the thirteen kinds of gobies which inhabit our marine waters are covered in this book: for information on the other ten, please refer to the checklist and references in the appendices.

Meaning of name. – *Typhlogobius:* blind *Gobius; californiensis:* from California.

Hemiramphidae (Halfbeak Family)
Ribbon Halfbeak
Euleptorhamphus longirostris (Cuvier, 1829)

Distinguishing characters. – The halfbeaks have a beak only on the lower jaw, which sets them apart from other marine fishes in our area. The ribbon halfbeak is easily recognized by its very elongate body, a beak which is more than twice as long as its head, and pectoral fins which are also twice as long as its head.

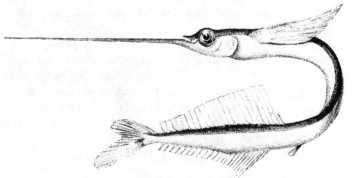

Fig. 37. *Euleptorhamphus longirostris*

Natural history notes. – *Euleptorhamphus longirostris* is found throughout all warm oceans of the world. In the eastern Pacific, it ranges from Newport Bay, California, to Peru and the Galapagos Islands. Maximum length has been reported as being "about 18 inches" (45.7 cm) (North America) and "19 inches" (48.3 cm) (New Guinea), but we measured a female dipnetted near Magdalena Bay in February 1959 that was 19-3/4 inches (49.8 cm) long. Another ribbon halfbeak collected off Nicaragua was 18

71

inches long (45.6 cm), not counting the beak or tail fin, so the species probably attains 23 or 24 inches (58.4 to 60.9 cm) at least.

Much of their life is spent in oceanic waters, but they move inshore to spawn. In the eastern north Pacific this inshore movement takes place during spring, but we have no information on spawning season. The eggs have long sticky tendrils and are demersal, in that they sink to the bottom, where they adhere to the substrate or to objects growing on the substrate.

Ribbon halfbeaks, as with other halfbeaks, are capable of "flight," and there are several accounts of *E. longirostris* gliding as far as 50 feet (15.2 m) without additional propulsion. Their remains have been found in the stomachs of several kinds of oceanic predators, but we have no information on halfbeak food items.

The genus has been reported as a fossil in the Californian Miocene, based upon a skeleton imprinted in diatomite, but we doubt if this is correctly identified. Halfbeak otoliths found in Miocene deposits near Bakersfield do not seem to represent *Euleptorhamphus* either.

Capture data. – Ribbon halfbeaks are attracted to bright lights suspended above the water's surface at night, and most individuals are dipnetted when so attracted. Sometimes they are caught with small mesh encircling nets set for other species, but usually they pass through the mesh of these nets or jump over the float line.

Other family members. – Three other kinds of halfbeaks have been found or caught off California on one or two occasions each. They are fairly easy to recognize by checking one or two characters:

1. If the pectoral fins and beak are both more than twice as long as the head, it is a ribbon halfbeak.
2. If the pectoral fins are only about as long as the head, the beak is less than half again as long as the head, and
   a. the pelvic fin insertion is closer to the base of the tail than to the pectoral base, it is a longfin halfbeak.
   b. the pelvic fin insertion is closer to the gill opening than to the caudal base, and
      i. the mandible has a red tip and its length projects 1.2

72

to 1.6 times into head length, it is a California half-beak.

    ii. the mandible lacks a red tip and projects only 0.8 to 1.3 times into the head, it is a common halfbeak.

Meaning of name. – *Euleptorhamphus:* very slender beak; *longirostris:* long beak.

Hexagrammidae (Greenling Family)
Painted Greenling
Oxylebius pictus Gill, 1862

Distinguishing characters. – The very typical body shape and coloration serve admirably to distinguish this small fish. The sharply tapered snout, single lateral line, and 12 to 13 soft anal rays are not characteristic of other greenlings, and the gray to light brown body with vertical dark red bars precludes mistaking a painted greenling for anything else. Males usually have much black coloring showing through, including the bars.

Fig. 38. *Oxylebius pictus*

Natural history notes. – *Oxylebius pictus* ranges from the Queen Charlotte Islands, British Columbia, to Point San Carlos, Baja California. It seldom strays far from rocky habitat and has been taken intertidally and to depths of 186 feet (56.7 m). It has been reported to reach a length of 10 inches (25.4 cm), but the largest of more than 500 individuals measured over a ten-year period was less than

73

7 inches (17.3 cm). This fish weighed about 2-1/2 ounces (70 g) and was eight years old; a ripe female 6 inches long (15.4 cm) weighed just over 2 ounces (62 g) and was five years old.

Spawning takes place from February until November, at least, and the cluster of orange-colored eggs adheres to low-growing algae on a rocky substratum. The males are thought to guard the nest until hatching occurs, because during this period they are very pugnacious and will charge all intruders, including skindivers. We have no information on number of eggs per spawning, number of spawnings per season, or incubation period.

Painted greenlings have a preference for a wide assortment of crustaceans and polychaete worms. Occasionally small mollusks and bryozoans are also eaten. Remains of painted greenlings have been found in stomachs of bass and rockcod; they probably provide fodder for other predatory fishes in their environment also.

Capture data. – Painted greenlings are strictly demersal, and are easily caught alive by scuba divers. They adapt well to aquarium life and make a colorful and interesting addition. Most records for the species, however, are based upon innumerable specimens collected by scientific personnel using ichthyocides in the rocky subtidal zone. There are a few records of painted greenlings being caught on tiny hooks baited with shrimp, and sometimes they are caught in lampara nets or purse seines which have been set near rocky habitat for other species.

Other family members. – Some ichthyologists place painted greenlings in their own family (Oxylebiidae), while others combine painted greenlings, combfishes, (Zaniolepidae), lingcod (Ophiodontidae) and other greenlings (Hexagrammidae) into one large family, Hexagrammidae. We prefer to keep the combfishes in their own family, but have combined the others. For further information, please refer to the checklist and references in the appendices and to our accounts in *Marine Food and Game Fishes of California.*

Meaning of name. – *Oxylebius:* sharp *Lebius* – *Lebius* being a junior synonym (i.e., a duplicate name and therefore incorrect) for *Hexagrammos,* thus a hexagrammid with a sharp snout; *pictus:* painted.

74

Kyphosidae (Sea Chub Family)
Halfmoon
Medialuna californiensis (Steindachner, 1875)

Distinguishing characters. — This is the only perch-shaped fish in our waters that is a uniform blue in color, dusky above and lighter below. The second dorsal and anal fins are composed of 22 to 27 and 17 to 21 soft rays respectively, and are solidly covered with a dense layer of scales.

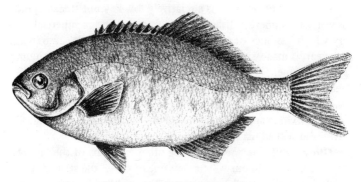

Fig. 39. *Medialuna californiensis*

Natural history notes. — *Medialuna californiensis* ranges from the Klamath River to the Gulf of California, and is also found offshore at Guadalupe Island. They usually form small loose schools in the mid-water column, especially in and around rocky habitat or where there is lush kelp, but at times they are solitary. Adults have been noted at all depths between the surface and 130 feet (39.6 m). Maximum reported size is 19 inches (48.3 cm) and 4 pounds and 12 ounces (2154 g).

Females are ripe from June through September, but spawning behavior, number of eggs per female, number of spawnings per season, and incubation period are unknown. After hatching the postlarval and young halfmoons are presumed to lead a pelagic existence. Often several hundred inch-long (2.54 cm) individuals can be found under floating patches of kelp and other flotsam many miles offshore. They move into inshore areas when less than

2 inches (5.1 cm) long, and mature at about 7-1/2 inches (19.1 cm) when in their second year. Maximum age appears to be about eight years, but may be a year or two longer in exceptional cases.

Their principal food appears to be algae — browns, reds, and greens — but they will also eat fair quantities of bryozoans, sponges, and other invertebrates, especially encrusting, sedentary, or semisedentary types. We have no information on halfmoon predators, but they should be excellent fare for some of the pinnipeds that live in their habitat.

Capture data. — There is a relatively low-key but steady sport as well as commercial fishery for halfmoons. The commercial fishery which has once or twice exceeded 50,000 pounds (22,675 kg) per year, is made exclusively with nets. Sportfishermen, including spearfishermen, account for an estimated 67,000 halfmoons per year, mostly south of Point Conception. These probably average one-half pound (226 g) each. Pelagic juveniles an inch or two (2.5 to 5.1 cm) long are easily dipnetted alive from under floating kelp rafts, and will adjust to aquarium life quite rapidly.

Other family members. — Three other members of the family have been taken in our waters, but one of these, the striped sea chub, is quite rare. All can be easily distinguished by their colors:

1. If the fish is dark olive-green, has blue eyes, and one or two cream-colored spots on the back at the base of the dorsal fin, it is an opaleye.
2. If it is dusky blue above and lighter blue below, and has 22 to 27 soft dorsal rays, it is a halfmoon.
3. If the body is dusky-brown, yellowish-brown, or light-gray, there are fewer than 15 soft dorsal rays, and
   a. there are a dozen or more dark vertical bars on each side, it is a zebraperch.
   b. there are dark spots on the scales which form longitudinal stripes on each side, it is a striped sea chub.

Meaning of name. — *Medialuna:* half moon; *californiensis:* California.

Labridae (Wrasse Family)
Rock Wrasse
Halichoeres semicinctus (Ayres, 1859)

Distinguishing characters. – This is the only buck-toothed fish in our waters which has large firmly-fixed body scales, a dorsal fin composed of 9 fairly sharp spines and 11 or 12 soft rays, a lateral line which drops abruptly beneath the end of the dorsal fin, and body coloring primarily greenish-brown. Adults all have red eyes, and the male has a blue-black band on his side behind a yellow pectoral fin. Young rock wrasses, up to about 4 inches (10.2 cm), are an overall orange-brown and have two white streaks running the length of the fish.

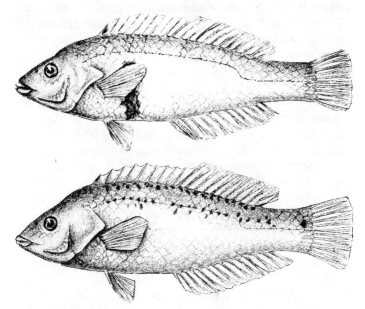

Fig. 40. *Halichoeres semicinctus:* upper, male; lower, female

Natural history notes. – *Halichoeres semicinctus* ranges from Point Conception, California, to the Gulf of California. It is not uncommon, but usually is alone when observed. Their typical habitat is one of rocky substratum along the outer coast, especially areas of dense vegetation. They have been taken from the shallow intertidal zone up to depths of 78 feet (23.8 m). Maximum reported length is 15 inches (38.1 cm), but the largest verifiable length was 14 inches (35.6 cm). This fish, a male, weighed 1 pound 10 ounces (727 g) and was nine years old, judged by growth zones on its otoliths.

Females are ripe during summer months and presumably spawn then. The eggs are believed to be pelagic, but we have no information on number per female, incubation period, number of spawnings per female per year, or other intimate details of reproduction. The tiny young are first seen in September. Females apparently first mature when they are two years old, and since they are protogynous hermaphrodites they undergo a sex change and become males when about five years old. Thus, all large rock wrasses are males.

When startled or badly frightened, a rock wrasse may dive into the sand and completely bury itself. They often sleep buried in the sand at night, with only the head protruding. Their food generally has been noted as crustaceans, mollusks, bryozoans, algae, and serpent stars. We have no information on the predators which eat *Halichoeres,* however.

There is a skeletal imprint and bony remains of a fossil rock wrasse from the Miocene Altimira shales on the Palos Verdes Peninsula; it may not be *H. semicinctus.*

Capture data. – An estimated 2400 rock wrasse are caught by hook-and-line fishermen or speared each year, almost all south of Santa Barbara. Perhaps a hundred or so are taken accidentally in lamparas and similar nets which are set near rocky habitat for other species. They are not considered a desirable food fish, but small individuals are highly prized for the color and interest they add to a marine aquarium. They are fairly easily captured alive by experienced scuba divers.

Other family members. — The three wrasses which inhabit our waters are easily distinguished at all sizes. All, of course, have protruding jaw teeth.

1. If there are 12 spines in the dorsal fin, it is a California sheephead.
2. If it is cigar-shaped, orange-brown, with a large black blotch at the base of the tail, it is a señorita.
3. If neither of the above fits, it is a rock wrasse.

Meaning of name. — *Halichoeres:* sea pig; *semicinctus:* half-banded.

Señorita
Oxyjulis californica (Günther, 1861)

Distinguishing characters. — This small cigar-shaped, buck-toothed wrasse has 9 or 10 dorsal spines, 13 dorsal soft rays and a lateral line that drops abruptly near the end of the dorsal fin, and is dusky-orange above and yellowish-orange below, with a dark chocolate spot on the caudal peduncle.

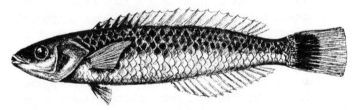

Fig. 41. *Oxyjulis californica*

Natural history notes. — *Oxyjulis californica* ranges from Sausalito, California, to Cedros Island, Baja California, but has not been seen north of Santa Cruz for a great many years. Señoritas are usually in loose schools or aggregations, but sometimes are solitary. They prefer mid-depths in the water column and have been taken in the shallow subtidal zone and in 180 feet (54.9 m) of water, but usually are in water shallower than about

79

75 feet (22.9 m). Maximum length, probably an estimate, is given as 10 inches (25.4cm); an 8-1/2-inch (21.6 cm) female weighed nearly 3 ounces (80 g) and was four years old, judged by growth zones on the otoliths.

Señoritas apparently mature when they are a year old, but we have no information on number of eggs or whether they spawn more than once each season. Spawning takes place from May through August, at least, and the eggs are said to be pelagic. When a señorita is frightened, it often will dive through kelp fronds and into the sand at the bottom. They often bury themselves in the sand at night to sleep with just their heads sticking out. Señoritas are notorious parasite pickers, removing an assortment of ecto-parasites from larger predatory fishes. In addition, they eat an assortment of small gastropods, crustaceans, worms, larval fish, and other items associated with kelp, including bits of seaweed as well. They do not feed off the bottom. Their remains have been found in stomachs of cormorants, bass, and a few other predators.

Otoliths of *O. californica* have been found in several Pliocene and Pleistocene deposits, and their jaws have been found in a coastal Indian midden.

Capture data. — An estimated 156 señoritas are caught on hook and line by sportfishermen each year, but this estimate seems a bit low to us. They have tiny mouths and are excellent bait thieves, but with small hooks a good fisherman can easily catch 20 or more in a few hours. They are not considered either desirable or edible. There is no commercial catch, but some are netted inadvertently in other fisheries. They adapt well to life in an aquarium, but are a bit large for most home aquaria.

Other family members. — For information on other Californian wrasses, please refer to the previous account on the rock wrasse.

Meaning of name. — *Oxyjulis:* sharp *Julis* — or, literally, a sharp-nosed fish which resembles *Julis,* an old world genus of wrasses; *californica:* Californian.

Liparididae (Snailfish Family)
Showy Snailfish
Liparis pulchellus Ayres, 1855

Distinguishing characters. — The soft, flabby, scaleless body, suction pad formed by the modified pelvic fins, tadpolelike head, and dorsal and anal fins which are joined to the caudal fin so that the fish tapers to a point posteriorly are sufficient characters to distinguish the showy snailfish. The body may be light to dark brown in color, but more often than not the sides and back are covered with wavy purplish lines which run the length of the fish, and sometimes are outlined in white.

Natural history notes. — *Liparis pulchellus* ranges from Peter the Great Bay, U.S.S.R., to Monterey Bay, California. It is an inhabitant of firm mud-bottom areas, and has reportedly been captured from the intertidal zone into depths of 300 feet (91.4 m) off Canada and 600 feet (182.9 m) off California. Although the maximum reported length is 10 inches (25.4 cm), the largest for which we can find an actual measurement was 8-1/2 inches (21.7 cm). An 8-inch (20.3 cm) fish was five years old, and several others which were slightly smaller were four.

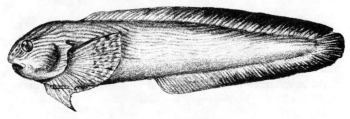

Fig. 42. *Liparis pulchellus*

Some showy snailfish apparently are mature when a year old and about 4 inches (10.2 cm) long. At that size a female contains around 900 eggs, while one 7 inches (17.8 cm) long will contain 9,000. Spawning takes place during winter and spring, but we

81

have no information regarding incubation time, whether a single female spawns more than once per season, or other facets or reproduction.

Their stomachs have been found to contain a variety of food items, including crustaceans, polychaete worms, gastropods, cumaceans, small flatfish, and even small individuals of their own kinds. Snailfish occasionally furnish a partial meal for predatory flatfishes and other denizens of mud- or sand-bottom habitat.

Capture data. – Fair numbers of showy snailfish are caught accidentally in shrimp trawls each year, but they have no commercial value so they usually are either thrown back or sold for use in pet food. They are not very hardy as aquarium fishes.

Other family members. – For information on the nineteen liparids which have been reported from waters off California, please refer to the checklist and references in the appendices.

Meaning of name. – *Liparis:* sleek-skinned; *pulchellus:* pretty.

Mugilidae (Mullet Family)
Striped Mullet
Mugil cephalus Linnaeus, 1758

Distinguishing characters. – The striped mullet is easily distinguished from all other fishes by its two widely separated dorsal fins, forked tail, anal fin with 8 or 9 soft rays, and small terminal V-shaped mouth. Atherinids (jacksmelt and relatives) might be mistaken for mullet the first time one sees them, but they have 22 or 23 soft rays in the anal fin.

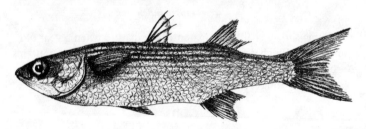

Fig. 43. *Mugil cephalus*

Natural history notes. – *Mugil cephalus* has been reported as ranging from about Monterey Bay, California, to Chile, but in recent years has seldom been seen north of Playa del Rey. At times mullet have been very abundant in the Salton Sea, where they have provided flourishing commercial and sport fisheries. They are most commonly observed in coastal estuaries, bays, lagoons, and flowing waterways (irrigation canals and such) throughout the Imperial Valley and from Los Angeles County south. They aggregate in small schools comprising a dozen to several hundred individuals; if they are present in a given body of water, a few individuals will be seen jumping at the surface at almost any time of day. They are reported to attain lengths of 3 feet (91.4 cm), but the largest noted in the Salton Sea during investigations of the fishery were 2 feet long (60.9 cm) and weighed just under 10 pounds (4.5 kg). Many of these large fish were sixteen years old, which is about maximum for the species.

M. cephalus apparently spawns during winter months in the eastern Pacific, and it is generally believed that spawning takes place well offshore over deep water. Silvery postlarvae and juveniles are often captured in plankton nets and with dipnets near the surface, many miles at sea. At just over an inch (2.5 cm) in length they commence moving shoreward, and 2- to 5-inch (5.2 to 12.7 cm) individuals are abundant in shallow water. Mullet up to a foot (30.5 cm) long or longer commonly enter freshwater streams and rivers, and some individuals have been known to travel fifty miles (80.5 km) or more upstream. Female mullet attain larger sizes than males.

The pelagic larvae are eaten by numerous surface-feeding predators including tropical tunas, whereas juveniles to 6 inches (15.2 cm) or so are preyed upon rather heavily by fish-eating birds (pelicans, gulls, terns, frigate birds, etc.). Mullet food habits have been studied in many parts of the world, and diatoms, blue-green algae, green filamentous algae, cladocerans, and other microscopic plants and animals comprise the bulk of their diet.

Mullet otoliths have been found in 25-million-year-old Miocene deposits near Bakersfield, but these do not appear to be from *M. cephalus*. Otoliths have also been found in Indian middens near the Salton Sea.

83

Capture data. – Mullet seldom will take a baited hook, but in irrigation canals and other freshwater habitat where they are abundant, they sometimes can be caught with a tiny hook that has been baited with a small fragment of earthworm, or with a tiny dry fly. When mullet were abundant in the Salton Sea, fish were commonly snagged with a surf-casting outfit and a large treble hook while they aggregated in shallow water near the mouths of various freshwater inlets. In conjunction with this method, clubs, spears, and dipnets were also used effectively.

Commercial fishing has been almost entirely with gill nets, primarily in the Salton Sea and the salt ponds in southern San Diego Bay. The fishery has been sporadic at both localities because of variations in mullet abundance, gear restrictions, closed seasons, poor market conditions, and other factors. In the past fifty years the commercial catch has ranged from about 1300 pounds in 1941 to a peak of 503,000 pounds in 1945. The flesh is white but quite oily; it makes a topnotch smoked product.

Other family members. – No other member of the family is known from California, although at least seven other mullets exist in the tropical eastern Pacific.

Meaning of name. – *Mugil:* to suck, in reference to the mullet's presumed feeding behavior; *cephalus:* pertaining to the head, which in the mullet is quite distinctive.

Mullidae (Goatfish Family)
Mexican Goatfish
Mulloidichthys dentatus (Gill, 1862)

Distinguishing characters. – Although goatfishes are readily recognized by their very typical shape, large scales, and generally quite colorful bodies, the two fleshy chin whiskers which are used in food detection dispel any uncertainty as to their identification.

Natural history notes. – *Mulloidichthys dentatus* has been recorded from Long Beach, California, to Tres Marias Islands, Mexico, and possibly northern Peru, but their capture off Encinitas and Long Beach in 1919 appears to be the only report of the species north of about Magdalena Bay. They have been taken in shallow subtidal areas and into depths of 125 feet (38.1 m), at least. Maximum reported length is one foot (30.5 cm), but this may be an estimate rather than an actual measurement. They are a schooling species, and often aggregate with other than their own kind.

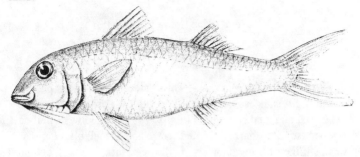

Fig. 44. *Mulloidichthys dentatus*

The Mexican goatfish is reported to be a deep red in color with some horizontal yellow streaks along its sides. The small specimens taken off Long Beach and Encinitas were "marked with crimson and yellow bands."

Although we have no information on reproduction or on ages, their food habits have been noted. They feed primarily at night and eat a variety of invertebrates and some fishes. Most abundant items include polychaete worms, mollusks, benthic crustaceans, and small stargazer fishes. They locate their prey by sweeping the bottom with their sensitive chin whiskers. Juvenile goatfishes, including *M. dentatus,* are often found in surface waters many miles at sea, and these inch to two-inch-long (2.5 to 5.1 cm) youngsters are eaten at every opportunity by tunas, jacks, dolphinfish, and some marine mammals which inhabit the pelagic realm in tropical waters.

Capture data. – The only specimens known from California were caught with bottom trawls. Where goatfishes are abundant, they can be and are captured with gill nets and beach seines, and many are speared. They make very interesting aquarium specimens because of their colors as well as behavior, but they generally are too large for any but public aquaria.

Other family members. – No other goatfish is known within several hundred miles of California.

Meaning of name. – *Mulloidichthys: Mullus*-like fish, alluding to a genus of goatfishes found in other parts of the world; *dentatus:* toothed.

Muraenidae (Moray Family)
California Moray
Gymnothorax mordax (Ayres, 1859)

Distinguishing characters. – This is an eel which lacks pectoral and pelvic fins; it has a leathery skin and low fleshy ridges where the dorsal and anal fins would be. Its jaws are filled with well-developed, serrate, sharply pointed teeth, and the gill opening is a small slitlike opening on a level with the mouth, but well behind it.

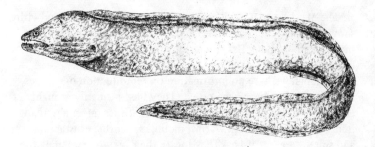

Fig. 45. *Gymnothorax mordax*

Natural history notes. – *Gymnothorax mordax* ranges from north of Santa Barbara to Santa Maria Bay, Baja California, and is especially common around the rocky shores of our southernmost offshore islands. Though secretive during daylight hours, they can be found in almost any rocky area to depths of about 50 feet (15.2 m) and have been reported as deep as 130 (39.6 m). They are reported to attain a length of 5 feet (152.4 cm), which they

86

unquestionably do, but the largest we have measured was a 47-1/2-inch-long (120.7 cm) male that weighed 14-1/2 pounds (6.6 kg).

In a recent study, otoliths were used to determine moray ages, and 12- to 16-inch (30.5 to 40.6 cm) fishes were found to be two or three years old. At 30 inches (76.2 cm) they are five or six years old, and a 43-incher (190.2 cm) was twenty-two. They apparently live longer than this, because one was kept in an aquarium for twenty-six years before it died, and it was several years old when placed in the aquarium. Spawning is thought to take place during summer, but we have no information on fecundity, age at first spawning, spawning behavior, egg size or shape, hatching time, or even whereabouts of the larvae. In fact, young morays smaller than about 8 or 10 inches (20.3 to 25.4 cm) are unknown.

Their favorite food appears to be fish, mostly blacksmiths which retire at night into rocky holes, the time and place where morays are most active. They also feed heavily on embiotocid perch, shrimp, crabs, octopuses, and squid. A large moray speared at San Clemente Island had eaten two foot-long (30.5 cm) flying-fishes.

Teeth from *G. mordax* have been found in coastal Indian middens and some Pleistocene and Pliocene deposits in southern California.

Capture data. – Shorefishing at night in shallow rocky habitat, especially at Santa Catalina, will yield excellent catches of morays. Poke-pole fishing during daylight hours also is productive. Spearfishermen who are willing to live dangerously take fair numbers each year, but they seldom utilize the morays they spear. Commercial lobstermen usually trap more than are caught by all other methods, but almost without exception these are thrown back alive. They make extremely interesting aquarium pets, but are just a bit too large for the average home.

Other family members. – No other member of the family is known within several hundred miles of California.

Meaning of name. – Gymnothorax: naked breast, alluding to the lack of scales; *mordax:* prone to bite.

Ophichthidae (Snake Eel Family)
Spotted Snake Eel
Ophichthus triserialis (Kaup, 1856)

Distinguishing characters. – The stiff, pointed tail, which is never reached by either the dorsal or anal fin, and the dusky-yellowish body, adorned with a wide assortment of black spots and dots, will distinguish the spotted snake eel from all other fishes in our waters.

Natural history notes. – *Ophichthus triserialis* ranges from Eureka, California, to Peru. It generally inhabits intertidal and shallow subtidal areas, but has been taken up to depths of 60 feet (18.3 m), and probably goes much deeper. Maximum reported size is 44-1/2 inches (113.5 cm) and 4-1/2 pounds (1935 g).

Very little information is available regarding the life history of the spotted snake eel, even though large numbers are caught by shrimp trawlers in the Gulf of California. Several females caught off southern California in July, August, and September have been full of very large, apparently mature eggs, so their spawning season includes the late summer months, at least. A 32-inch-long (81.3 cm) male which weighed 1-1/4 pounds (567 g) was nine years old, judged by some excellent growth zones on its otoliths.

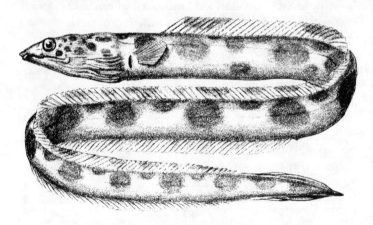

Fig. 46. *Ophichthus triserialis*

88

Larval snake eels are elongate, semitransparent, boneless, gelatinous creatures known collectively as leptocephali, and they often comprise a fair portion of the fish catch made with fine-mesh midwater trawls many miles offshore. We do not know how long they remain as leptocephali, but presumably within a year or two they transform into their adult shape and take up an existence on or in the substrate. Snake eels generally burrow into the sand tail-first, but once buried they can travel in either direction.

Clam parts, small fish, and shrimp have been found in the stomachs of spotted snake eels, but they undoubtedly eat many other items also. We have no information regarding species which prey upon them, however.

Capture data. – Most adult spotted snake eels have been caught in fine-mesh bottom trawls or on hook and line. Although they are not abundant north of about Magdalena Bay, Baja California, perhaps 20 individuals have been taken in local waters, including the record-sized 44-1/2-incher (113.5 cm). They are extremely hardy and live well in an aquarium, but because of their large size and burrowing habits (rarely in sight), they are not desirable for the average home aquarium.

Other family members. – Two other members of the family have been taken off California, but all are rather easy to identify:

1. If the body is covered with black spots, it is a spotted snake eel.
2. If the body is plain-colored and
   a. the dorsal fin starts above the pectoral fin and ends before reaching the tail, it is a yellow snake eel.
   b. the dorsal fin starts well behind the pectoral fin and connects with the tail fin, it is a Pacific worm eel.

Meaning of name. – *Ophichthus:* snake fish; *triserialis:* three-rowed.

Ophidiidae (Cusk-eel Family)
Basketweave Cusk-eel
Otophidium scrippsi Hubbs, 1916

Distinguishing characters. – The eellike body, and the pair of chin whiskers (modified pelvic fins) near the tip of the lower jaw will distinguish the cusk-eel among our marine fishes. The lack of spots on the body will separate the basketweave cusk-eel from the deeper-dwelling spotted cusk-eel. Regardless of their eellike appearance, cusk-eels are more closely related to the cods than to the true eels.

Natural history notes. – *Otophidium scrippsi* ranges from Point Arguello, California, to Guaymas, in the Gulf of California. They have been taken at depths between about 9 and 230 feet (2.7 to 70.1 m), but are most numerous at depths shallower than 60 feet (18.3 m). The maximum reported length is 10-3/4 inches (27.3 cm); a 10-incher (25.4 cm) weighed only 5 ounces (141.7 g), so we doubt that maximum weight would exceed one-half pound (226.7 g).

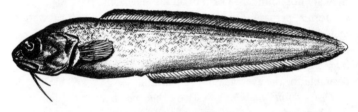

Fig. 47. *Otophidium scrippsi*

We have no information on life history except a note that small crustaceans have been found in their stomachs. They undoubtedly eat many other food items, however, because several have been caught on hooks baited with fish and clam. They are believed to be mostly nocturnal in their habits, but do come out on dark or dismal days, or when the water is murky. When startled or disturbed, they will burrow tail-first into sandy substrate, or move backward into a hole or crack.

They appear to be a choice food item among sea lions, and are eaten by a variety of large predatory fishes and a few birds such as cormorants.

Their otoliths have been found in several Pliocene and Pleistocene deposits.

Capture data. – Occasional individuals are caught on hook and line, and even more rarely one is speared. They are taken incidental to commercial fisheries utilizing encircling nets or trawl gear.

They can be caught by scuba divers using handheld nets, and readily adjust to aquarium life. Many are sucked into the cooling water system of coastal electric steam generating plants — these do not survive, of course.

Other family members. — The two cusk-eels are easily distinguished on the basis of color:

1. If the body is spotted, it is a spotted cusk-eel.
2. If the body is plain-colored, it is a basketweave cusk-eel.

Meaning of name. — *Otophidium:* alluding to the large saccular otolith plus *Ophidium,* the family name; *scrippsi:* in honor of Ellen W. Scripps.

Osmeridae (Smelt Family)
Surf Smelt
Hypomesus pretiosus (Girard, 1854)

Distinguishing characters. — The very typical body shape, dorsal fin at about midlength with anal fin directly below, presence of an adipose fin, and lack of concentric striations on the gill cover are sufficient to distinguish surf smelt from all other fishes in our waters.

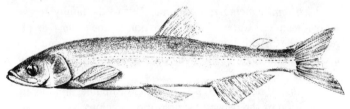

Fig. 48. *Hypomesus pretiosus*

Natural history notes. — *Hypomesus pretiosus* ranges from Prince William Sound. Alaska, to Long Beach, California, but is not common south of about San Francisco. It is strictly a marine species, seldom moving any great distance offshore, but sometimes straying into brackish water. Early reports give a maximum length of 12 inches (30.5 cm), but authenticated maxima appear to be 10 inches (25.4 cm) off California, and 8-3/4 inches (22.2 cm) in Canadian waters.

They spawn in the surf zone during daylight hours, and although some spawning appears to occur throughout the year, peak spawning is from May through October. Some females will spawn at age one, and all will spawn when two years old. Maximum age appears to be four years, but few fish exceed three. Depending upon size (age) a female will have 1500 to 30,000 eggs per spawning, and may spawn three to five or more times per season. The eggs are adhesive and stick to sand grains in the surf zone, where they hatch in ten to eleven days. There are six or eight very good spawning beaches in California, the southernmost being near Davenport.

Surf smelt feed mostly upon small crustaceans, but also eat polychaete worms, larval fish, jellyfish, and other items. They are often important in the diets of salmon, halibut, and other large predatory species.

Capture data. – Sportfishermen in California catch more numbers and pounds of surf smelt than any other single species. A-frame and two-man jump nets are used in the breaking surf during spawning runs. Surf smelt average about ten per pound (5 per kilogram), and in good years sportfishermen will take an estimated 400,000 pounds (181,400 kg). The take would be even higher, except for a daily bag limit of 25 pounds (11.3 kg) per licensed fisherman.

Other family members. – Six members of the family are known from our waters. For information on the other five, please refer to the checklist and references in the appendices and to our account of the night smelt in *Marine Food and Game Fishes of California.*

Meaning of name. – *Hypomesus:* below middle, alluding to the position of the pelvic fins; *pretiosus:* precious.

Ostraciontidae (Boxfish Family)
Spiny Boxfish
Ostracion diaphanum Bloch and Schneider, 1801

Distinguishing characters. – The peculiar shape of this small fish, with its body entirely encased in a hard, bony, boxlike covering, will distinguish it from all other fishes in our waters.

92

1. High tide at White Point, Los Angeles County. Note dirty water caused by waves pounding against cliffs.

2. Low tide six hours later at White Point. Fishes must be well adapted for living in this environment.

3. High tide near Point Arena in northern California.

4. Exposed kelp and rocks at low tide near Point Arena.

1. Bluebanded goby
   ready to dart into
   protective crevice
   from its resting place
   a few inches away.

2. Moray surrounded by
   cleaning shrimp.
   Note juvenile black-
   eye goby at side of
   cave and juvenile
   blacksmiths in front
   of moray.

3. Blackeye goby in
   typical pose.

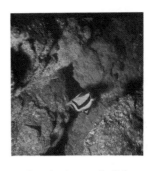

4. Scythe butterflyfish
   staying close to pro-
   tective caverns.

1. Stripefin ronquil resting on flat rocky bottom.

2. Snubnose sculpin at home in empty mussel shell.

3. Male convictfish or painted greenling on irregular rock bottom. Purple coralline algae covers rocks in foreground.

4. Male rock wrasse surrounded by black-eye gobies.

1. Grass rockcod in typical pose. Orange streak by head is mantle of a rock scallop. Pinkish-red patches above and below are a small form of anemone.

2. Whitebelly rockcod.

3. Honeycomb rockcod.

4. Gopher rockcod.

1. Aggregation of large opaleyes. Note long-spined sea urchins on bottom and giant kelp in background.

2. Small school of half-moons in midwater column near a rocky shore.

3. A C-O turbot resting on substrate of broken shells and pebbles.

4. A scorpionfish flying beside a gorgonian, with its head into the current.

1. Crevice kelpfish well camouflaged by its protective coloration.

2. Yellowfin fringehead (red phase) in typical pose and habitat: with head sticking out of empty piddock clam burrow.

3. Head of a mosshead warbonnet sticking out of empty tubeworm shell. A typical home and pose for this small fish.

4. Giant kelpfish.

TYPICAL INHABITANTS OF THE ROCKY SUBTIDAL

1. Island kelpfish.

2. Juvenile garibaldi. The blue spotting and blotching disappears as the fish gets older.

3. Sailfin sculpin with saillike dorsal fin held parallel to brown bar on side of head.

4. Notchbrow blenny.

1. Black-and-yellow rockcod and female convictfish sharing the same resting place.

2. Northern ronquil resting on its pelvic fins while keeping a watchful eye for danger.

3. Rock prickleback removed from nearby tidepool to photograph.

4. Tidepool snailfish found clinging to kelp and placed on rock to photograph.

Fig. 49. *Ostracion diaphanum*

Natural history notes. – *Ostracion diaphanum* is found in all
warm seas of the world; in the eastern Pacific it ranges from
Santa Barbara, California, to Peru. They generally are found at or
near the surface, sometimes where the water is several thousand
feet (several hundred meters) deep, but have been noted 90 feet
(27.4 m) beneath the surface. Maximum length is reported as
being 10 inches (25.4 cm) by some authors, and "about a foot"
(30.5 cm) by others, the largest from our waters was a 7-incher
(17.8 cm).

No information is available regarding spawning or reproduction,
ages, food habits, and other intimate facets of spiny boxfish life
history. We can only speculate that spawning occurs during
spring and summer, based upon the abundance of tiny juveniles
observed and collected throughout the eastern north Pacific
tuna-fishing grounds during late fall and winter months. We have
examined several stomachs, but all were empty. Our efforts to
determine ages have been fruitless.

At times, spiny boxfish comprise a fair portion of skipjack,
yellowfin tuna, and wahoo diets. They have also been found in
yellowtail and dolphinfish stomachs.

Capture data. – Most spiny boxfish which stray north into
our waters probably have been affected by the colder temperatures, because almost all have been found on various beaches
where they had drifted or been cast ashore by the waves. In more
southerly waters they are often abundant, and many have been dipnetted at the surface, caught in purse seines, or spit up by tunas
which were caught by commercial fishermen. They adjust very well

to life in an aquarium, but the water must be kept warm for best results. First reported from California in 1932, perhaps 20 individuals have been found here since.

Other family members. – No other member of the family lives within many hundreds of miles (or kilometers) of California.

Meaning of name. – *Ostracion:* shell; *diaphanum:* transparent.

Pholididae (Gunnel Family)
Saddleback Gunnel
Pholis ornata (Girard, 1854)

Distinguishing characters. – The unique body shape, a dorsal fin composed entirely of spines, and an anal fin which commences well back on the body (closer to the base of the tail than to the snout) will distinguish members of the gunnel family. The saddleback gunnel is characterized by U-shaped or V-shaped saddle marks on the back, tiny pelvic fins comprised of one spine and one ray, and 34 to 38 anal fin rays.

Fig. 50. *Pholis ornata*

Natural history notes. – *Pholis ornata* ranges from Peter the Great Bay, U.S.S.R., through the Bering Sea and south to Carmel Beach, California. They are primarily inhabitants of the intertidal and shallow subtidal zones, but have been taken as deep as 120 feet (36.6 m). Maximum reported length is 12 inches (30.5 cm).

The saddleback gunnel is found most often on muddy bottom areas where there are patches or beds of eelgrass, surf grass, or seaweeds. An examination of their otoliths indicates that they mature after their second winter and probably do not live much beyond seven years, if that long. Spawning is said to take place during late winter and spring, and both sexes are reported to guard the eggs until they hatch.

94

They feed upon a variety of small crustaceans and shelled mollusks. We have no information on saddleback gunnel predators, however.

Capture data. – Few saddleback gunnels are taken by other than fishery biologists or scientific personnel who are making general fish collections or studying the biota of a given area. They sometimes are stranded in weed beds and shallow mudlfat pools when the tide goes out, and at such times they may be picked up by curious strollers and tideflat scroungers.

Other family members. – Although six other gunnels are found in our waters, all are fairly easy to distinguish. The following simplified key should serve to identify all seven of these fishes:

1. If both pectoral and pelvic fins are absent, it is a kelp gunnel.
2. If the pectoral fins are present but the pelvics are absent, and
   a. the anal spine is deeply grooved on the outer surface, it is a penpoint gunnel.
   b. the anal spine is round (never grooved), it is a rockweed gunnel.
3. If both the pectoral and pelvic fins are present, and
   a. there are 34 to 38 anal rays and the markings on the back are V-shaped, it is a saddleback gunnel.
   b. there are 34 to 37 anal rays and the markings on the back are ( ) shaped, it is a crescent gunnel.
   c. there are 40 to 44 anal rays, it is a red gunnel.
   d. there are 50 to 53 anal rays, it is a longfin gunnel.

Meaning of name. – *Pholis:* one who lies in wait; *ornata:* ornamented.

Kelp Gunnel
Ulvicola sanctaerosae Gilbert & Starks, *in* Gilbert, 1897

Distinguishing characters. – The kelp gunnel differs from the other members of the family in that it lacks both the pectoral and pelvic fins. It usually is an overall yellowish-brown color, but may be red-brown or tan, depending upon the color of the kelp in which it lives.

Natural history notes. – *Ulvicola sanctaerosae* ranges from Pacific Grove, California, to Papalote Bay, south of Ensenada,

Baja California, and is also found at Guadalupe Island. They are inhabitants of the kelp canopy, and some of the kelp beds where they are found are growing in 125 feet (38.1 m) of water. Maximum reported length is 11-1/4 inches (28.6 cm).

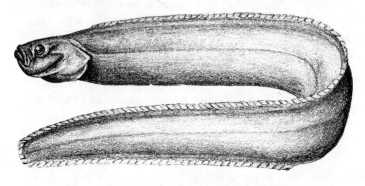

Fig. 51. *Ulvicola sanctaerosae*

Females are ripe during late winter, and presumably spawn then, as the young are present from March through June. Nothing is known about spawning habits or behavior, and to our knowledge no one has counted eggs in the ovaries of a mature kelp gunnel. An examination of a number of otoliths indicates that kelp gunnels mature after their second winter, and they probably do not live beyond eight years, if that long. Otoliths of an 8-1/2-inch-long (21.6 cm) fish had five growth zones on them.

They feed almost exclusively upon mysids and amphipods (crustaceans) which they find in and under the kelp canopy. Rarely, kelp gunnels are eaten by kelp bass or kelp-dwelling rockcods.

Capture data. — Until the early 1940s, kelp gunnels were considered rare, but about that time a study of the fish species which were hauled aboard barges during kelp-cutting operations revealed that they were abundant in the kelp canopy. They also are attracted to bright lights at night, and can be dipnetted at the surface when they swim to such a light suspended from the side of a drifting or anchored vessel.

Other family members. — For information on other family

members, please refer to the previous account on the saddleback gunnel and to the checklist in Appendix I.

Meaning of name. – *Ulvicola:* sea lettuce inhabitant; *sanctaerosae:* for Santa Rosa Island, the locality of first capture.

Pleuronectidae (Righteye Flounder Family)
Diamond Turbot
Hypsopsetta guttulata (Girard, 1856)

Distinguishing characters. – The eyed right side of this flatfish is gray-green to gray-black and profusely covered with round blue spots. It is readily distinguished from other right-eyed flatfishes by its small mouth, lack of dorsal fin rays on the blind side, and porcelain-white underside, except for lemon-yellow coloring around the mouth and near the dorsal fin origin.

Natural history notes. – *Hypsopsetta guttulata* has been recorded at many localities between Cape Menodcino, California, and Magdalena Bay, and in the northern Gulf of California. They

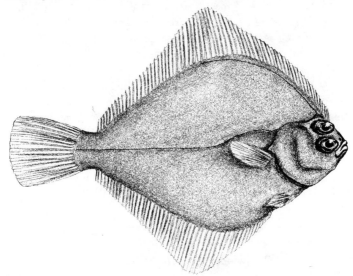

Fig. 52. *Hypsopsetta guttulata*

97

prefer quiet back bay, lagoon, and slough waters, but are found
in sheltered outer-coast habitat and have been taken as deep as 150
feet (45.7 m) but prefer depths shallower than about 15 feet (4.6
m). Although reported to attain a length of 18 inches (45.7 cm),
the largest of more than 1200 captured during trawling operations
in waters off southern California was 15 inches (38.1 cm) long
and weighed slightly less than 2 pounds (900 g). Specimens which
are 12 to 15 inches (30.5 to 38.1 cm) long appear to be eight or
nine years old, judged by growth zones on their otoliths.

Females become mature when two or three years old and
spawning takes place during spring, summer, and early fall, at
least. The eggs are said to drift in the pelagic realm until they
hatch, but information is lacking on number of eggs per adult
female, time required to hatch, etc.

The numerous stomachs which we have examined have contained polychaete worms, clam parts, and ghost shrimp almost
exclusively. We know of no predators of diamond turbots, but
feel certain that one occasionally falls prey to a hungry electric
ray or angel shark, both common inhabitants of the same habitat.

Capture data. — A recent survey revealed that some 5700 diamond turbots are caught by sportfishermen south of Point Conception each year. A few others are speared by novice skindivers,
and an unknown number is netted in shallow water incidental
to lampara and purse-seine fisheries.

Other family members. — For information on the other nineteen kinds of right-eyed flounders known from Californian waters,
please refer to the checklist and references in the appendices and
to the family accounts in *Marine Food and Game Fishes of
California.*

Meaning of name. — *Hypsopsetta:* deep flounder; *guttulata:*
with small spots.

Pomacentridae (Damselfish Family)
Garibaldi
Hypsypops rubicundus (Girard, 1854)

Distinguishing characters. — This is the only perch-shaped
(laterally compressed) fish in our waters which as an adult is uni-

formly golden-orange in color. Juveniles are red-orange with iri-
descent blue blotching and spots. The deeply-forked tail fin,
with rounded dorsal and ventral lobes, and very large body scales
are also quite distinctive.

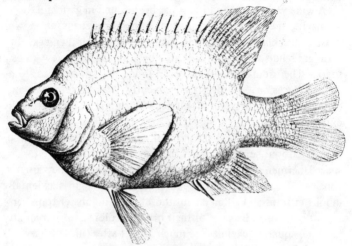

Fig. 53. *Hypsypops rubicundus*

Natural history notes. – *Hypsypops rubicundus* ranges from
Monterey Bay, California, to Magdalena Bay and offshore at
Guadalupe Island, but is not abundant north of about Santa
Barbara. They inhabit rocky bottom areas, both on the exposed
outer coast and in quiet semiprotected or protected areas of shallow bays. Garibaldis have been noted as deep as 95 feet (28.9 m),
but most are in depths shallower than 45 (13.7 m). Maximum
length is reported as 15 inches (38.1 cm), but few are seen which
are longer than a foot (30.5 cm). They mature when about five
years old and are reported to live to perhaps seventeen, although
an adult male at Steinhart Aquarium was said to have been twenty-
nine years old when it died.

Garibaldis breed from March through July, and the female
deposits her eggs in a nest which the male has prepared by cleaning
off all but calcareous items. He then allows only three kinds of
red algae to grow in the center of the cleared area, and the eggs
are laid on these. A single female will lay between 15 and 80
thousand eggs, depending upon her size, and some nests contain

up to 190 thousand eggs, obviously produced by several females. The male remains at the nest site and defends it against all intruders except sea lions. The eggs hatch in two or three weeks, depending upon water temperatures.

A wide variety of food items has been found in garibaldi stomachs, including sponges, sea anemones, bryozoans, algae, worms, crustaceans, clams and mussels, snail eggs, and their own eggs. In turn, garibaldis sometimes are eaten by sea lions and a few kinds of sharks. Their remains have also been found in giant sea bass stomachs on occasion. Señoritas are frequently observed removing and eating parasitic crustaceans (fish lice) which infest adult garibaldis externally.

Capture data. – Garibaldis are fully protected by law, because their territorial behavior makes them extremely vulnerable to spearfishermen and could result in their decimation. They are not considered very choice as table fare. A few are caught accidentally in lobster traps and gill nets, but those found in lobster traps can be returned alive. Because of their brilliant colors and adaptability to home aquaria, juveniles are much sought after (illegally) by scuba divers using slurp guns.

Other family members. – The blacksmith is the only other damselfish known in our waters. As an adult it has a dusky-blue body, profusely covered with black spots posteriorly. Juvenile blacksmiths are purplish-blue anteriorly and yellowish posteriorly.

Meaning of name. – *Hypsypops:* high below eye, for the wide preorbital region; *rubicundus:* red.

Pomadasyidae (Grunt Family)
Salema
Xenistius californiensis (Steindachner, 1875)

Distinguishing characters. – The laterally-compressed body (perch-shaped), large eye, and 6 to 8 brassy or orange-brown stripes along the sides of the body will distinguish the salema from other fishes in our waters. The grunting noises made by these fishes when removed from the water will clinch their identification.

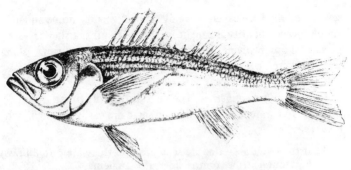

Fig. 54. *Xenistius californiensis*

Natural history notes. — *Xenistius californiensis* ranges from Monterey Bay, California, to Peru, but is not common north of about Point Dume, California. They school in loose aggregations comprising perhaps ten to several hundred individuals, and are found in a wide variety of habitats (rocky, sandy, exposed shore, sheltered bays) throughout their range. Their preferred depths appear to be from about 4 to 35 feet (1.2 to 10.7 m). Maximum reported length is 10 inches (25.4 cm), but this may be an estimate. The largest of several thousand which we have seen was 8-1/2 inches long (21.8 cm) and weighed about 4 ounces (113 g).

Spawning takes place during spring and early summer, and the eggs are reported to be pelagic. We have no information on number of eggs per female, number of spawnings per season, time needed for the eggs to hatch, age at first maturity, or maximum age. Tiny juveniles are often observed schooling with sargo and black croaker juveniles of similar sizes. In fact, juveniles of these three species are very similar in shape and coloration. In an aquarium, salemas will reach a length of about 3-1/2 inches (8.9 cm) in a year.

They are preyed upon by numerous fish, mammal, and bird species which live in or travel through their habitat. We have not examined salema stomachs to see what they feed on, however.

Otoliths of *X. californiensis* have been found in several Pleistocene fossil deposits in southern California.

Capture data. — An estimated 5,000 salemas are caught each year south of Point Conception by sportfishermen, and perhaps several hundred are caught alive by scuba divers for home aquaria. An unknown quantity is caught incidental to other fisheries, especially those involving the use of lampara nets. When tuna fishing was conducted with hook and line, salemas were an important bait species.

Other family members. – The sargo is the only other grunt found in our waters. The salema and sargo are easily distinguished however:

1. If the two dorsal fins are separate, and there are 6 to 8 brassy horizontal stripes on the sides, it is a salema.
2. If the two dorsal fins are broadly joined, and there is a dusky vertical bar midlength of the fish, it is a sargo.

Meaning of name. – *Xenistius:* strange sail, in allusion to the dorsal fin; *californiensis:* Californian.

Priacanthidae (Bigeye Family)
Catalufa
Pseudopriacanthus serrula (Gilbert, 1890)

Distinguishing characters. – This is the only perch-shaped fish in our waters which has a crimson-red body with several darker red vertical bars (most noticeable when alive or freshly caught) along the sides, and an extremely large eye. The pelvic fins are black to dusky on their tips, especially in young fish.

Natural history notes. – *Pseudopriacanthus serrula* ranges from Monterey Bay, California, to Talara, Peru, and is also found at the Galapagos and Revillagigedos Islands. They live at or near the bottom in rocky habitat, and appear to be mostly nocturnal. They have been captured or observed at depths of 30 to 198 feet (9.1 to 60.4 m) but probably are found both shallower and deeper. The largest we have seen was a 13-inch (33.0 cm) female which weighed 2.2 pounds (997 g) and was fifteen years old, judged by growth zones on its otoliths. This and an 8-1/2 incher (21.6 cm) were caught in gill nets near Magdalena Bay, Baja California, in May 1970.

We have no information on reproduction or reproductive behavior. The few stomachs we have examined have contained mostly fish and crustacean remains.

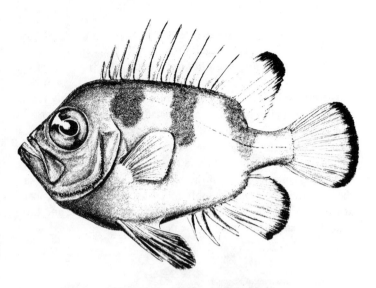

Fig. 55. *Pseudopriacanthus serrula*

Some authorities have placed this fish in genus *Pristigenys* because of a presumed resemblance to an Eocene fossil from Europe. However, the holotype of this fossil was imperfect and few details of the head and shoulder-girdle are clearly preserved. Extremely long pelvic fins are visible on the fossil, however, and these in conjunction with the associated fish fauna leads us to believe that *Pristigenys* is a senior synonym (i.e., first recorded name) of *Cookeolus,* rather than *Pseudopriacanthus. Cookeolus,* a large-eyed priacanthid, has extremely long pelvic fins and is found almost exclusively in a pelagic environment well away from rocky shores.

Capture data. – Although not over a dozen individuals have been noted in our waters, they have been captured in gill nets, traps, and trawl nets, speared, caught on hook and line, and photographed alive. Scuba divers report that catalufas are unafraid and can be picked up by hand when encountered in their natural surroundings.

Other family members. – No other member of the family is known within several hundred miles (or kilometers) south of California.

Meaning of name. – Pseudopriacanthus: false *Priacanthus; serrula:* a little saw.

103

Sciaenidae (Croaker Family)
Black Croaker
Cheilotrema saturnum (Girard, 1858)

Distinguishing characters. — The black croaker is readily recognized by its typical body shape (especially head and shoulder region) and color, especially if one is familiar with its close relatives. When freshly caught, the creamy to brown or blackish body has purplish overtones, and there may or may not be a light or dark bar about midlength of the fish. There is a jet-black edge on the gill cover, the mouth is subterminal, there is no chin whisker at the tip of the lower jaw, the preopercle is not serrate, and there is no black spot at the base of the pectoral fin.

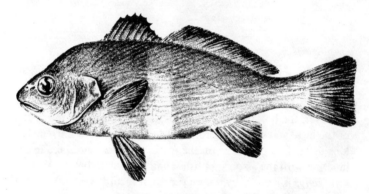

Fig. 56.   *Cheilotrema saturnum*

Natural history notes. — *Cheilotrema saturnum* ranges from Point Conception to Magdalena Bay. They school in loose aggregations of up to fifty or more individuals, and are found at all depths between the surface and about 150 feet (45.7 m) especially where the bottom is rocky. Maximum authenticated length is 15 inches (38.1 cm), although early accounts state that they attain 18 inches (45.7 cm). They are quite secretive and hide in caves, rocky caverns, and crevices during the day. At night they often forage over sandy or muddy bottoms, but retire to their rocky retreats at dawn.

Females mature when they are two or three years old and about 9 inches (22.7 cm) long. Spawning takes place during spring and summer, and the eggs, which are pelagic, are reported to hatch in two or three days. The tiny juveniles frequently school with similar-sized salemas and sargos, to which they bear a striking resemblance. A 14-1/2-inch (36.8 cm) fish weighed 1 pound 9 ounces (708 g), but its age was not determined. Maximum age appears to be at least twenty years.

Black croakers feed exclusively on crustaceans, mostly crabs and shrimp. In turn, black croakers are eaten by a few kinds of elasmobranchs, sea lions, and an occasional giant sea bass.

Capture data. — An estimated 5700 black croakers are caught by sportfishermen in southern California each year, including those speared by scuba divers. Incidental catches are made with gill nets which have been set for other species, but electric steam-generating plants make greater inroads on the black croaker population each year than do sport and commercial fisheries combined.

Other family members. — For information on the other seven kinds of croakers known from our waters, please refer to the checklist and references in the appendices and to our accounts of this family in *Marine Food and Game Fishes of California.*

Meaning of name. — *Cheilotrema:* lip pore; *saturnum:* dusky, saturnine.

Scorpaenidae (Scorpionfish Family)
China Rockcod
Sebastes nebulosus Ayres, 1854

Distinguishing characters. — The China rockcod is easily recognized by color alone. It is an overall blackish, mottled with yellow and white, but its most distinctive marking is a broad yellow stripe which runs from the tip of the third dorsal spine and along the membrane between that and the fourth spine down to the lateral line, and along that to the base of the tail.

Natural history notes. — *Sebastes nebulosus* ranges from southeastern Alaska to Diablo Cove, California, and San Miguel Island.

It has been caught or observed at depths of 36 to 420 feet (10.9 to 128 m) but usually is found shallower than 150 (45.7 m). Maximum reported length is 17 inches (43.2 cm), but we have no information on weight or age of this fish. A 12-1/2-inch-long (31.8 cm) female weighed 2 pounds (908 g), and was older than ten years.

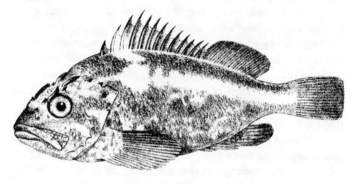

Fig. 57.   *Sebastes nebulosus*

As with other rockcods (genus *Sebastes*), fertilization is internal and embryos develop within the ovaries of the female. Spawning takes place during the winter, and depending upon size, each female will liberate many thousands of tiny embryos. The few that survive to age two will average about 5 inches (12.7 cm) in length and weigh between one and two ounces (28 to 56 g).

Food items most often observed in their stomachs are squid and octopi, crabs, shrimp and similar custaceans, and small fish. Remains of China rockcod have been noted in the stomachs of a lingcod, but we do not know of any other predators. Although rockcod otoliths are common in fossil deposits and coastal Indian middens, none of these has been identified as being from *S. nebulosus* as yet.

Capture data. – An estimated 4,000 China rockcod are caught by sportfishermen on hook and line along rocky stretches of coast north of Point Arguello each year. A few large adults are speared by scuba and skindivers. The commercial catch is mostly incidental to other fisheries, and of minor significance. They make interesting aquarium fishes, but outgrow a small space by the time they are a couple of years old.

Other family members. – For information on the 62 members of the scorpionfish family which have been reported from our waters (58 of these belong to genus *Sebastes*), please refer to the checklist and references in the appendices and to our accounts in *Marine Food and Game Fishes of California.*

Meaning of name. – *Sebastes:* magnificent; *nebulosus:* clouded.

Grass Rockcod
Sebastes rastrelliger (Jordan and Gilbert, 1880)

Distinguishing characters. – The mottled greenish-gray to blackish body, and thick caudal peduncle, are sufficient for recognizing this rockcod; but as a clincher, the rakers on the first gill arch are short and blunt, being barely longer than thick.

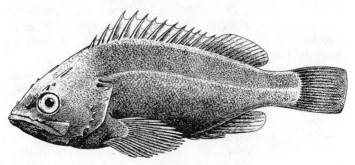

Fig. 58. *Sebastes rastrelliger*

Natural history notes. – *Sebastes rastrelliger* ranges from Yaquina Bay, Oregon, to Playa Maria Bay, Baja California. They are primarily inhabitants of rocky, wave-swept shores, but may also be found in more sheltered areas where the habitat is satisfactory. Although they have been taken from the intertidal zone and as deep as 150 feet (45.7 m), the species is rare at depths greater than about 60 feet (18.3 m). Maximum recorded length is 22 inches (55.9 cm). A 21-inch (53.3 cm) fish weighed 6 pounds 11 ounces (3032 g); maximum weight probably does not exceed 7 pounds (3174 g).

Grass rockcod mature when three or four years old and about 8 inches (20.3 cm) long; at this age and size they will weigh nearly one-half pound (226g). Spawning occurs during winter months, and a large adult will produce tens of thousands of tiny living embryos per spawning. A fish 16 inches (40.6 cm) long weighed 3.4 pounds (1541 g) and was fourteen years old, while an 18-3/4-inch (47.6 cm) female weighed just under 4-1/2 pounds (2040 g) and was sixteen or seventeen years old. Just prior to spawning, the enlarged ovaries in these large females often make up one-fifth or more of the fish's total weight.

A wide assortment of food items has been found in their stomachs: crabs, shrimp, octopi, snails, and small fish. Among these, crabs and octopi appear to be the preferred prey.

Capture data. – Sportfishermen are estimated to catch about 12,600 grass rockcod north of Point Arguello each year; and although no figures are available for southern California anglers, an equal number probably is taken by these fishermen. The commercial catch is negligible, a few being taken incidental to other fisheries. They are hardy and live well in an aquarium, but not many are kept in home aquaria because of their eventual large size.

Other family members. – Please refer to the previous account, on China rockcod.

Meaning of name. – *Sebastes:* magnificent; *rastrelliger:* a rake and to bear, alluding to the stubby gill rakers.

Scytalinidae (Graveldiver Family)
Graveldiver
Scytalina cerdale Jordan and Gilbert, 1880

Distinguishing characters. – Graveldivers are characterized by their lack of pelvic fins and scales, by a body which is deepest behind the vent, and by a rounded caudal fin which is joined to both the dorsal and anal fins. Their skin is quite loose, and the body color is an overall yellowish to pinkish.

Natural history notes. — *Scytalina cerdale* ranges from the Bering Sea to Diablo Cove, San Luis Obispo County, California. They burrow into coarse loose sand in intertidal and subtidal areas and into depths of 25 feet (7.6 m) or more. They may move about above the sand at night, but there are too few observations to justify stating this with certainty. Maximum reported length is 6 inches (15.2 cm).

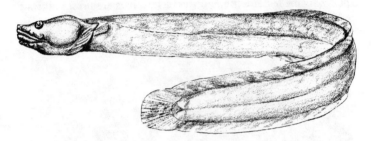

Fig. 59. *Scytalina cerdale*

We have no information on spawning season or behavior, and very little on other facets of their life history. A 3-inch-long (7.6 cm) fish was three years old, judged by an examination of its otoliths. Several individuals which were 5 to 5-1/4 inches (12.7 to 13.3 cm) long, appeared to be eight or nine years old, and had matured when they were three.

Capture data. — Few graveldivers are taken or seen by others than fishery, museum, or university personnel who are collecting with ichthyocides. Occasionally a few individuals will be dug out of the sand by clammers. They are quite hardy, but do not make good aquarium fishes because of their secretive habits.

Other family members. — There is no other member of family Scytalinidae.

Meaning of name. — *Scytalina:* viper, in allusion to its snake-like appearance; *cerdale:* wary one.

109

Serranidae (Sea Bass Family)
Spotted Sand Bass
Paralabrax maculatofasciatus (Steindachner, 1868)

Distinguishing characters. — This bass is distinguished by its
10 sharp, strong dorsal spines, the third one being the longest;
and by its densely spotted body. In juveniles the colors are quite
bright, and there are dusky horizontal stripes and broader vertical
bars. With age, the horizontal streaks eventually disappear and the
spots become less intense, but the vertical bars are present throughout life.

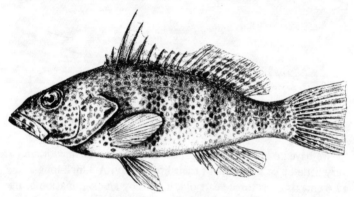

Fig. 60. *Paralabrax maculatofasciatus*

Natural history notes. — *Paralabrax maculatofasciatus* has been
reported as ranging from San Francisco, California, to Mazatlan,
Mexico, but it has seldom been seen north of Monterey, California, since the 1880s. They prefer sandy or muddy bottom
areas of flat relief from the shallow subtidal zone into water 200
feet (60.9 m) deep. Maximum reported size is 22 inches
(55.9 cm) and 5 pounds 10 ounces (2550 g). They probably attain
an age of twelve to fifteen years, at least; a 15-inch (38.1 cm)
fish was five years old, and an aquarium specimen lived for eight
years before dying — apparently not from old age.

Spotted sand bass spawn during spring and early summer, and
the eggs are reported to be pelagic. Unfortunately, we have no

information regarding number of eggs spawned per female, whether a single fish spawns more than once per year, or how long it takes the eggs to hatch.

Their food consists of crustaceans, small fish, and cephalopods primarily, but they feed upon many other forms which live in the same habitat. They probably are eaten by sea lions and a few other large predators, but we have no proof of this. We do know that juveniles are occasionally caught and eaten by cormorants.

Otoliths of *Paralabrax* have been found in fossil deposits of Pleistocene age and in coastal Indian middens, but it was not determined if they were from *P. maculatofasciatus.*

Capture data. – An estimated 12,000 spotted sand bass are caught by sportfishermen each year, exclusive of those taken from partyboats. No figures are available for the partyboat catch, but it is not believed to be significant. It has been against the law to take spotted sand bass (and other kinds of *Paralabrax*) commercially since 1947. Although they make an interesting and hardy addition to an aquarium, they soon outgrow the limited space of the average home aquarium.

Other family members. – For information on the other nine members of the family, please refer to the checklist and references in the appendices and to our accounts in *Marine Food and Game Fishes of California.* Some authorities place the giant sea bass and striped bass in a different family (Percichthyidae), but we prefer to retain them in the family Serranidae until interrelationships of the various basses are better understood.

Meaning of name. – *Paralabrax:* near or like *Labrax,* a European bass; *maculatofasciatus:* spotted and banded.

Sparidae (Porgy Family)
Pacific Porgy
Calamus brachysomus (Lockington, 1880)

Distinguishing characters. – The Pacific porgy is the only fish in our waters with a perch-shaped (laterally compressed) body, a broad scaleless area between the lips and the eye, a pectoral fin longer than the head, and a brownish-pink body (darker above) with some silvery-white shining through. There is a smoothly rounded heavy bony ridge above each eye, and the rear margin of the preopercle is raxor-sharp. The outline of a Pacific Porgy in side view is unmistakable. Juveniles have several chocolate-colored vertical bars on their sides, which are also present on adult porgies but fade almost as soon as the fish is hauled from the water.

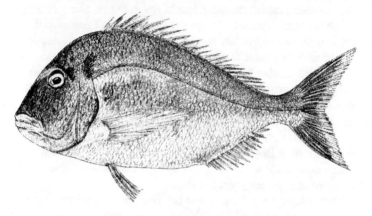

Fig. 61. *Calamus brachysomus*

Natural history notes. — *Calamus brachysomus* has been recorded from Oceanside, California (twice), to 150 miles (241 km) south of Lima, Peru, but it apparently does not spawn north of about Sebastian Viscaino Bay, Baja California. Pacific porgies probably reach a length of slightly less than 2 feet (60 cm) and a weight of 5 pounds (2267 g) or more but the largest one officially measured was 20 inches (50.8 cm) long; its weight is unknown. A 16-3/4-inch fish (42.5 cm) weighed 2-1/2 pounds (1135 g). Examination of more than a dozen sets of otoliths from large individuals indicates they attain an age of at least fifteen years.

Spawning apparently takes place during the spring months (based upon the presence of fish-of-the-year in mid summer),

112

and most individuals will spawn when they are three or four years old. We have no information on the number of eggs spawned per female, egg size, locality of spawning, or other facets of reproduction.

Young fish up to 4 or 5 inches (10.2 to 12.7 cm) long are abundant in very shallow water where the bottom is sandy, or firm sandy mud, particularly in coastal bays and lagoons. The adults also live over smooth (nonrocky) bottom areas, but they are usually found in slightly greater depths. They appear to be most abundant in 20 to 60 feet (6.1 to 18.3 m) of water, and there is one authenticated capture from 225 feet (68.6 m). Their natural food seems to be an assortment of sedentary or semi sedentary mollusks (clams and small snails) and crustaceans (shrimp and crabs). On rare occasions small fishes will be fed upon, but we have never observed more than one in a single stomach.

A large molar tooth (from the back part of the jaw) of *C. brachysomus* has been found in Pleistocene deposit at San Pedro, California, that was laid down about 120,000 years ago, when local ocean temperatures were much warmer than they are today.

Capture data. – The Pacific porgy has been caught only twice north of Mexico, so it cannot be deemed a sought-after sport species in our waters. Porgies readily take a baited hook, but they are excellent bait thieves, and the large, flattened molarlike teeth in each jaw make it difficult to hook one solidly. They will take a variety of baits including cut anchovy, but best catches are made with pieces of clam, mussel, shrimp, or squid.

Porgies have been brought in from Mexican waters and sold as "tai" in the fresh fish markets, but landings have been sporadic and light. In 1931, landings of nearly 1 ton (907 kg) were reported, but this figure has never been equaled since then.

Other family members. – No other member of the family is known within several thousand miles (or kilometers) of California.

Meaning of name. – *Calamus:* a quill or reed, for the quill-like interhaemal bone; *brachysomus:* short body.

113

Stichaeidae (Prickleback Family)
Mosshead Warbonnet
Chirolophis nugator (Jordan and Williams,
in Jordan and Starks, 1895)

Distinguishing characters. — The typical body shape, dorsal fin composed entirely of spines, and abundant plumelike or hairlike cirri on the top of the head are sufficient to distinguish the mosshead warbonnet from other fishes in our waters. The color pattern is also distinctive: both sexes are reddish-brown or orange-brown with vague dusky blotching or bars on the sides; males have 12 or 13 ocelli along the dorsal fin, while in the females these "eye spots" are replaced by dusky bars.

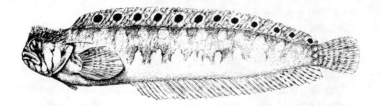

Fig. 62. *Chirolophis nugator*

Natural history notes. — *Chirolophis nugator* has been recorded from Kodiak Island, Alaska, to Point Arguello, California, on the mainland, and Cuyler Harbor, San Miguel Island. They are most abundant in the intertidal and shallow subtidal zones, but have been taken as deep as 264 feet (80.5 m). Maximum reported length is just over 5-1/2 inches (14.3 cm).

Mosshead warbonnets prefer living in and around hard rocky habitats. They frequently may be found living around bases of rocks, in crevices and tubeworm holes, and similar relief. When in a tube or narrow crevice, only the head protrudes, and the plumelike cirri above the eyes serve to camouflage them.

Spawning takes place during late winter and early spring and the large eggs — each about 2 mm in diameter — are presumed to adhere to the substrate. The adults may even guard their eggs until they hatch; but this is only speculation, based upon the be-

114

havior of other family members. We have no information on how long it takes their eggs to hatch, nor upon the larval or juvenile stages. An examination of otoliths and gonads indicates they first mature when they are two years old when they average about 4 inches (10.2 cm) in length and less than one-half ounce (12.5 g) in weight. A 5-1/4-inch fish (13.3 cm) weighed less than an ounce (18 g) and was 5 years old.

The only food we have found in their stomachs has been the remains of nudibranchs (sea slugs), including one kind, *Phidiana pugnax,* which uses toxic sting cells (nematocysts) as a defense mechanism. These sting cells apparently are swallowed alive by the nudibranchs, along with the coelenterates to which they belong, and are then incorporated into the cell-structure of the nudibranch itself. Since they are supposed to function in the nudibranch as in the coelenterates which produce them originally, it would be interesting to know why they do not affect the mosshead warbonnet, and if the fish can or does make use of them similar to the nudibranch. The cirri on the mosshead warbonnet's head resemble very much the respiratory apparatus (cerata) of some nudibranchs. We do not know who preys upon them.

Capture data. – Mosshead warbonnets are probably never seen nor captured by other than fishery biologists or museum and university personnel making scientific collections with the aid of ichthyocides, or by an unusually observant scuba diver.

Other family members. – For information on the thirteen kinds of pricklebacks which are known in our waters, please refer to the checklist and references in the appendices and to our account of the monkeyface prickleback in *Marine Food and Game Fishes of California.*

Meaning of name. – *Chirolophis:* hand comb; *nugator:* a top.

Crisscross Prickleback
Plagiogrammus hopkinsii Bean, 1893

Distinguishing characters. – The dorsal fin of this blue-black or gray-black fish is comprised entirely of spines, but its most distinctive character is the two lateral lines which branch vertically, form-

115

ing platelike divisions or cells on the fish's sides. A chocolate-brown stripe extends from the tip of the snout onto the body, as in some other pricklebacks. These stripes are more prominent in juveniles; they tend to be brown in the juveniles and blue-black in adults.

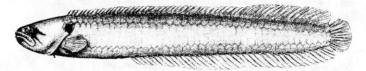

Fig. 63. *Plagiogrammus hopkinsii*

Natural history notes. – Plagiogrammus hopkinsii has a very limited range within California – from Pacific Grove to Point Arguello and San Nicolas Island. They inhabit rocky stretches along the outer coast where there is dense algal cover, and have been taken from the intertidal zone into 70 feet (21.3 m) of water. Maximum reported length is 7-3/4 inches (19.7 cm).

An examination of their gonads indicates that they spawn during spring and summer, but actual spawning has not been observed. Some appear to mature when two years old and all are mature at three, when they average about 4 inches (10.2 cm) long. A 6-1/4-inch fish (15.7 cm) weighed about 1-1/4 ounces (35 g) and was six years old. Based upon this observation, maximum age apparently does not exceed ten years.

The stomachs which we have examined have contained an assortment of invertebrates but mostly crustaceans (mysids, amphipods, and shrimp) and mollusks (chiton remains).

Otoliths of *P. hopkinsii* were found in an Indian midden at Diablo Cove – presumably they arrived there in the stomach of some predator which had eaten the crisscross prickleback before the Indian caught it.

Capture data. – Few of these small fishes are seen or captured by other than fishery biologists or museum and university personnel making scientific collections with the aid of ichthyocides.

Other family members. – Please refer to the previous account, on the mosshead warbonnet.

Meaning of name. – *Plagiogrammus:* oblique line; *hopkinsii:* for Timothy Hopkins, founder of the Seaside Laboratory at Pacific Grove — now Hopkins Marine Station, Stanford University's marine laboratory.

Rock Prickleback
Xiphister mucosus (Girard, 1858)

Distinguishing characters. – The rock prickleback is evenly colored greenish-gray or greenish-black, except for two darker bars which diverge backward and downward from the rear of the eye. The dorsal fin is composed entirely of spines, there are no pelvic fins, and there is but one anal spine. It differs from its nearest relative, *X. atropurpureus,* in having the origin of the dorsal fin above the pectoral fin. In *X. atropurpureus,* the dorsal fin commences at least one headlength behind the pectoral fin.

Fig. 64. *Xiphister mucosus*

Natural history notes. – *Xiphister mucosus* ranges from Port San Juan, Alaska, to Point Arguello, California. They inhabit rocky outer-coast areas which have a dense algal cover, and in this habitat they have been found at depths from the intertidal zone to 60 feet (18.3 m). Maximum known length is 23 inches (58.6 cm).

Rock pricklebacks are mature when five years old and about 14 inches (35.6 cm) long. Spawning occurs during the winter, and the relatively large eggs probably adhere to the substrate until they hatch. Although we did not make any egg counts, the ovaries of a 19-inch (48.3 cm) female which was ready to spawn and weighed just over a pound (472 g) comprised nearly one-ninth of the fish's total weight. Females appear to grow larger than the

117

males, and may live longer. The largest fish we have examined for age was a 20-1/2-inch (52.1 cm) female; it turned out to be eleven years old. The heaviest fish we have seen was a 20-1/4-inch (51.4 cm) male which tipped the scales at 2 pounds (923 g).

Rock pricklebacks feed almost exclusively on algae; more than a dozen kinds of seaweed have been identified from their stomachs. Other food items include an occasional shrimp, bryozoan, small fish, or piece of sponge.

Judged by the otoliths found in an Indian midden at Diablo Cove, rock pricklebacks apparently were important in the diet of these primitive peoples.

Capture data. – Fewer than one hundred rock pricklebacks are caught by sportfishermen in our coastal waters each year. Mostly they are collected by fishery biologists, museum personnel, and university teachers or students who are studying intertidal and subtidal faunas or conducting special research on the species.

Other family members. – Please refer to the account of the mosshead warbonnet.

Meaning of name. – *Xiphister:* a small sword; *mucosus:* slimy.

Stromateidae (Butterfish Family)
Pacific Pompano
Peprilus simillimus (Ayres, 1860)

Distinguishing characters. – The metallic silvery laterally compressed (perchlike) body, lack of pelvic fins, deeply forked tail fin, and long pectorals will distinguish the Pacific pompano from all other species in our waters.

Natural history notes. – *Peprilus simillimus* ranges from British Columbia to about Abreojos Point, Baja California. It usually inhabits shallow water near shore, and often forms small, but fairly dense, schools. They are reported to reach a length of 10 inches (25.4 cm), but few individuals have been seen that exceed 8 (20.3 cm). A female that was just over 8-1/2 inches (21.6 cm) long weighed slightly less than 5 ounces (141 g), and appeared to be just past four years old.

118

Fig. 65.   *Peprilus simillimus*

Spawning takes place during every month of the year, and the eggs are believed to be pelagic, but no valid information is available on the subject. They are believed to feed upon small crustaceans, but their food habits have not been studied. Small pompano have been found in the stomachs of California halibut, barracuda, and kelp bass; they are undoubtedly eaten by numerous other predators also.

To date, their remains have not been found in Indian middens or fossil deposits, although *Peprilus* otoliths have been found in Atlantic coast Miocene. (Four species presently live off the Atlantic and Gulf coasts of the United States.)

Capture data. – Sportfishermen catch an estimated 10,000 Pacific pompano each year, mostly from piers, and mostly with snagging gear. They are a highly prized food fish, but the commercial catch, made almost entirely with encircling nets, is insignificant compared with numerous other Californian fisheries. The several electric steam-generating plants which have intake systems for their cooling waters on the open coast probably make greater inroads on the Pacific pompano population than do the sport and commercial fisheries combined.

Other family members. – No other member of the family occurs in our waters, although two or three species are known between about Magdalena Bay and Panama.

Meaning of name. – *Peprilus:* derived from the Greek, meaning one of Hesychian's unknown fish; *simillimus:* very similar – in reference to an Atlantic species.

Syngnathidae (Pipefish and Seahorse Family)
Kelp Pipefish
Syngnathus californiensis Storer, 1845

Distinguishing characters. — A pipefish sometimes is alluded to
as a seahorse in which the head failed to turn down and the tail
didn't curl up — but this isn't entirely true. Pipefishes also have
caudal fins (which are lacking in seahorses), and the brood pouch
is a covered slit on the belly of the male (in the seahorse it is
sealed, except for a small opening in the belly beneath the dorsal
fin). The kelp pipefish is distinguished from other native pipefishes
by its larger size, 17 to 22 body rings, and 44 to 50 tail rings.

Natural history notes. — *Syngnathus californiensis* ranges from
San Francisco, California, to Santa Maria Bay, Baja California.
They are inhabitants of the kelp beds, usually near the canopy,
and occasionally are found near or in floating patches of kelp many
miles offshore. Their color usually matches the algal forests with
which they associate, and may be yellowish, brownish, or greenish.
Maximum reported length is 19-1/2 inches (49.5 cm). A 17.8-inch
male (45.4 cm) weighed just one ounce (27 g).

Fig. 66. *Syngnathus californiensis*

The eggs are fertilized as the female deposits them in the brood
pouch of the male. This breeding routine usually transpires over a
several-day period, as only a few eggs are transferred at a time, with
the female resting between each batch. In our waters, "pregnant"
males have been observed from September through December,
but we lack information on the time needed for kelp pipefish
eggs to hatch. In some pipefishes the eggs will hatch in eight to ten
days, but for others it may take twice as long.

Pipefishes feed upon tiny planktonic crustaceans such as
mysids, small shrimp, and amphipods. They can (and do) suck

these into the mouth from a distance of an inch (2.5 cm) or more; this takes place so rapidly that it is nearly impossible for the human eye to follow the action. In turn, pipefishes have been found in a variety of fish stomachs, including tunas, and they are a choice food item of gulls.

Pipefish skeletal imprints and remains have been found in numerous fossil deposits in southern California which are of Miocene age. They are especially abundant in some diatomites and shales, but none of these fossils, to our knowledge, is *S. californiensis*.

Capture data. – The best place to collect pipefishes is on kelp-cutting barges, because they are drawn aboard these vessels with the cut kelp. Occasionally pipefishes are cast shore with clumps of kelp torn up by storms, and they are often attracted to a bright light suspended above the water at night. Some adults are occasionally swept out to sea either by currents, storms, or with drifting patches of kelp. They are easily collected by scientific personnel using ichthyocides, and scuba divers have no difficulty catching them with handheld nets, if they can see them to begin with. They make extremely interesting aquarium pets and will readily accept brine shrimp.

Other family members. – There is disagreement among various ichthyologists as to the number of genera and species of syngnathids inhabiting our waters, and some even split the species into subspecies. We follow a recent publication (see Miller and Lea in our list of references in Appendix III) in our decision to distinguish only four pipefishes and one seahorse at this time:

1. If the caudal fin is absent, the head is at a right angle to the body, and the brood pouch of the male is sealed except for a porelike opening, it is a Pacific seahorse.
2. If there is a caudal fin, the head is on the same axis as the body, the brood pouch of the male is an elongate, covered slit, and
   a. there are 14 to 15 body rings, 20 to 23 soft dorsal rays, and a double row of spots along the sides of the body, it is a snubnose pipefish.
   b. there are 14 to 15 body rings, 26 to 34 soft dorsal rays, and no spots on the sides, it is a barred pipefish.
   c. there are 17 to 22 body rings, 36 to 46 tail rings, and 28

121

to 44 soft dorsal rays, it is a bay pipefish.
d. there are 17 to 22 body rings, 44 to 50 tail rings, and 36 to 47 soft dorsal fin rays, it is a kelp pipefish.

Meaning of name. – *Syngnathus:* together jaws; *californiensis:* Californian.

Tetraodontidae (Puffer Family)
Bullseye Puffer
Sphoeroides annulatus (Jenyns, 1842)

Distinguishing characters. – The typical body shape, rabbitlike teeth, dorsal and anal fins set well back on the body, and pattern of dusky rings on the back, in the form of a bullseye, will distinguish this puffer from all other fishes in our waters. In addition, the skin may be smooth or have minute prickles, and the interorbital width, when measured into snout length, goes less than three and a half times. The fish gets its common name by its ability to swallow water or air and puff up like a balloon – elsewhere they are called swellfishes, balloonfishes, and similar names.

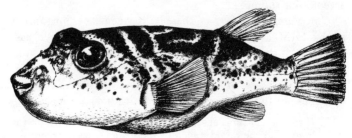

Fig. 67. *Sphoeroides annulatus*

Natural history notes. – *Sphoeroides annulatus* is reported as ranging from San Diego, California, to the Galapagos Islands and Peru, but it is extremely rare north of about Magdalena Bay, Baja California. They are primarily inhabitants of shallow nearshore areas, and especially abundant in quiet coastal waters and protected bays, but occasionally are found well offshore where they have been carried by some erratic current. Maximum reported

122

length is "about 15 inches" (38.1 cm), but this appears to be an estimate only.

We have no information on their feeding or reproductive behavior. We do know from observation that they sometimes sleep at night while wedged against some part of the substrate, and that during daytime a skindiver wearing swimfins can often outswim and catch puffers by hand because they get excited and stop to swallow water, which puffs them to such an extent that they can't swim.

Sphoeroides teeth have been found in some Miocene deposits in California, but these have not been associated with *S. annulatus* as yet.

Capture data. – Bullseye puffers can be caught on hook and line, as they readily accept such baits as cut fish or squid. Most, however, are taken with nets incidental to other fisheries. Sometimes they are cast ashore by storm waves, and they are easy to spear or catch by hand. There appear to be only two records from California, one in 1857 and the other sometime prior to 1880. The second record is quite vague, however, and may actually refer to the specimen or specimens collected at San Diego in 1857.

In most waters of the world puffer flesh is quite toxic, and many persons have died from eating it. In the western North Atlantic, on the other hand, a puffer forms the basis for a lively commercial fishery and its flesh is highly prized for food. To our knowledge, *S. annulatus* has not been implicated as a toxic species, but its greatest value appears to be as a curio – blown up, dried, and used as a den ornament of some type. Small puffers make extremely interesting and hardy aquarium pets, although they will chew the fins of other fishes.

Other family members. – Three puffers are known from our waters, but only the oceanic puffer has been found here on more than one or two occasions:

1. If there are 12 to 15 dorsal and anal rays, and the fish is blue-black above and white below (spots and bars are also present), it is an oceanic puffer.
2. If there are 6 to 8 dorsal and anal rays, the color is not as above, and
   a. the interorbital is shallowly concave to convex

123

and goes into snout length less than three and a half times, it is a bullseye puffer.
   b. the interorbital is deeply concave and goes more than four times into snout length, it is a longnose puffer.

Meaning of name. – *Sphoeroides:* sphere resemblance; *annulatus:* ringed.

Trichodontidae (Sandfish Family)
Pacific Sandfish
Trichodon trichodon (Tilesius, 1811)

Distinguishing characters. – This distinctively shaped little fish is grayish or brown above and white below, has a nearly vertical mouth, with a fleshy fringe on the lips, and very large pectoral fins, and lacks scales.

Natural history notes. – *Trichodon trichodon* ranges from Medni Island on the east coast of the Bering Sea to San Francisco, California. Throughout this range they inhabit sandy or muddy bottom areas of flat relief, and have been found from the intertidal zone into depths of 180 feet (54.8 m). Maximum reported length is 12 inches (30.5 cm) but this may be an estimate. A 10-1/2-inch long specimen (25.9 cm), the largest we could authenticate, weighed about one-half pound (223 g) and was eight or nine years old, judged by growth zones on its otoliths.

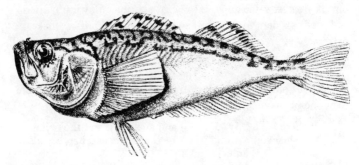

Fig. 68.   *Trichodon trichodon*

The Pacific sandfish spawns during winter months, and apparently matures when two years old, although many individuals do not mature until three. We have no information on the number of eggs spawned, nor the time they require to hatch. Their food is unknown, but *Trichodon* remains have been found in the stomachs of salmon and some kinds of flatfish, at least, and at times they are eaten by seals and sea lions.

Capture data. – There is no fishery for Pacific sandfish, and we know of no one who has eaten one. Fair numbers are caught incidental to trawl fisheries for shrimp and flatfishes; and where they are abundant, they can be caught by hand intertidally. They have been and are exhibited in aquaria, but because of their burrowing habits they do not make a good display.

Other family members. – No other member of the family is known in our waters.

Meaning of name. – *Trichodon:* hair tooth; *trichodon:* repetition of generic name with same meaning.

Appendix I
Checklist of Tidepool and Nearshore Fishes of California

(Asterisk indicates species description is given in text.)

Agonidae (Sea poachers)
Agonomalus species	Kelp poacher
Agonopsis emmelane (Jordan and Starks, 1895)	Northern spearnose
Agonopsis sterleyus (Gilbert, *in* Jordan and Evermann, 1898)	Southern spearnose
Agonus acipenserinus Pallas, *in* Tilesius, 1813	Sturgeon poacher
Anoplagonus inermis (Günther, 1860)	Smooth alligatorfish
Bathyagonus infraspinatus (Gilbert, 1904)	Spinycheek starnose
Bathyagonus nigripinnis Gilbert, 1890	Blackfin starnose
Bathyagonus pentacanthus (Gilbert, 1890)	Bigeye starnose
Bothragonus swanii (Steindachner, 1876)	Rockhead*
Ganoideus vulsus (Jordan and Gilbert, 1880)	Beardless spearnose
Occella verrucosa (Lockington, 1880)	Warty poacher
Odontopyxis trispinosa Lockington, 1880	Pygmy poacher
Pallasina barbata aix Starks, 1896	Tubenose poacher
Stellerina xyosterna (Jordan and Gilbert, 1880)	Pricklebreast poacher
Xeneretmus latifrons (Gilbert, 1890)	Blackedge poacher
Xeneretmus leiops Gilbert, 1915	Smootheye poacher
Xeneretmus ritteri Gilbert, 1915	Flagfin poacher
Xeneretmus triacanthus (Gilbert, 1890)	Bluespotted poacher

Ammodytidae (Sand Lances)
Ammodytes hexapterus Pallas, 1814	Pacific sand lance*

Anarhichadidae (Wolffishes)
Anarrhichthys ocellatus Ayres, 1855	Wolf-eel*

Antennariidae (Frogfishes)
Antennarius avalonis Jordan and Starks, 1907	Roughjaw frogfish*

Apogonidae (Cardinalfishes)
Apogon guadalupensis (Osburn and Nichols, 1916)	Guadalupe cardinalfish*

Atherinidae (Silversides)
Atherinops affinis (Ayres, 1860)	Topsmelt*
Atherinopsis californiensis Girard, 1854	Jacksmelt
Leuresthes tenuis (Ayres, 1860)	California grunion

Aulorhynchidae (Tube-snouts)
  Aulorhynchus flavidus Gill, 1861   Tube-snout*

Balistidae (Triggerfishes)
  Balistes polylepis Steindachner, 1876   Finescale triggerfish*
  Xanthichthys mento (Jordan and Gilbert, 1882)   Redtail triggerfish

Bathymasteridae (Ronquils)
  Rathbunella hypoplecta (Gilbert, 1890)   Stripefin ronquil*
  Rathbunella species   Bluebanded ronquil
  Ronquilus jordani (Gilbert, 1889)   Northern ronquil

Batrachoididae (Toadfishes)
  Porichthys myriaster Hubbs and Schultz, 1839 Specklefin midshipman*
  Porichthys notatus Girard, 1854   Plainfin midshipman

Belonidae (Needlefishes)
  Strongylura exilis (Girard, 1854)   California needlefish*

Blenniidae (Combtooth blennies)
  Hypsoblennius gentilis (Girard), 1854)   Bay blenny
  Hypsoblennius gilberti (Jordan, 1882)   Notchbrow blenny*
  Hypsoblennius jenkinsi (Jordan and Evermann, 1896)   Mussel blenny

Bothidae (Lefteye flounders)
  Citharichthys sordidus (Girard, 1854)   Pacific sanddab
  Citharichthys stigmaeus (Jordan and Gilbert 1882) Speckled sanddab*
  Citharichthys xanthostigma Gilbert, 1890   Longfin sanddab
  Hippoglossina stomata Eigenmann and Eigenmann,
    1890   Bigmouth sole
  Paralichthys californicus (Ayres, 1859)   California halibut
  Xystreurys liolepis Jordan and Gilbert, 1880   Fantail sole

Brotulidae (Brotulas)
  Brosmophycis marginata (Ayres, 1854)   Red brotula
  Cataetyx rubrirostris Gilbert, 1890   Rubynose brotula
  Dicrolene species   Roughhead brotula
  Lamprogrammus niger Alcock, *in* Woodmason
    and Alcock, 1891   Paperbone brotula
  Oligopus diagrammus (Heller and Snodgrass,
    1903)   Purple brotula*

Chaetodontidae (Butterflyfishes)
  Chaetodon falcifer Hubbs and Rechnitzer,
    1958   Scythe butterflyfish*
  Chaetodon humeralis Günther, 1860   Threebanded butterflyfish

Clinidae (Kelpfishes)

Alloclinus holderi (Lauderbach, *in* Jordan and
   Starks, 1907) — Island kelpfish
Chaenopsis alepidota (Gilbert, 1890) — Orangethroat pikeblenny
Cryptotrema corallinum Gilbert, 1890 — Deepwater blenny
Gibbonsia elegans (Cooper, 1864) — Spotted kelpfish*
Gibbonsia erythra Hubbs, 1952 — Scarlet kelpfish
Gibbonsia metzi Hubbs, 1927 — Striped kelpfish
Gibbonsia montereyensis Hubbs, 1927 — Crevice kelpfish
Heterostichus rostratus Girard, 1854 — Giant kelpfish
Neoclinus blanchardi Girard, 1858 — Sarcastic fringehead
Neoclinus stephensae Hubbs, 1953 — Yellowfin fringehead
Neoclinus uninotatus Hubbs, 1953 — Onespot fringehead*
Paraclinus integripinnis (Smith, 1880) — Reef finspot*

Clupeidae (Herrings)

Alosa sapidissima (Wilson, 1811) — American shad
Clupea pallasii Valenciennes, *in* Cuvier and
   Valenciennes, 1847 — Pacific herring*
Dorosoma petenense (Günther, 1867) — Threadfin shad
Etrumeus teres (DeKay, 1842) — Round herring
Harengula thrissina (Jordan and Gilbert, 1882) — Flatiron herring
Opisthonema medirastre Berry and
   Barrett, 1963 — Middling thread herring
Sardinops caeruleus (Girard, 1854) — Pacific sardine

Cottidae (Sculpins)

Artedius corallinus (Hubbs, 1926) — Coralline sculpin
Artedius creaseri (Hubbs, 1926) — Roughcheek sculpin
Artedius fenestralis (Jordan and Gilbert, 1883) — Padded sculpin
Artedius harringtoni (Starks, 1896) — Scalyhead sculpin
Artedius lateralis (Girard, 1854) — Smoothhead sculpin
Artedius meanyi (Jordan and Starks, 1895) — Puget Sound sculpin
Artedius notospilotus Girard, 1856 — Bonehead sculpin
Ascelichthys rhodorus Jordan and Gilbert, 1880 — Rosylip sculpin*
Blepsias cirrhosus (Pallas, 1814) — Crested sculpin
Chitonotus pugetensis (Steindachner, 1876) — Roughback sculpin
Clinocottus acuticeps (Gilbert, 1896) — Sharpnose sculpin
Clinocottus analis (Girard, 1858) — Woolly sculpin*
Clinocottus embryum (Jordan and Starks, 1895) — Calico sculpin
Clinocottus globiceps (Girard, 1857) — Mosshead sculpin
Clinocottus recalvus (Greeley, 1899) — Bald sculpin

Enophrys bison (Girard, 1854)	Buffalo sculpin
Enophrys taurina Gilbert, 1914	Bull sculpin
Hemilepidotus hemilepidotus (Tilesius, 1811)	Red Irish lord
Hemilepidotus spinosus (Ayres, 1854)	Brown Irish lord
Icelinus burchami Evermann and Goldsborough, 1907	Dusky sculpin
Icelinus cavifrons Gilbert, 1890	Pit-head sculpin
Icelinus filamentosus Gilbert, 1890	Threadfin sculpin
Icelinus fimbriatus Gilbert, 1890	Fringed sculpin
Icelinus oculatus Gilbert, 1890	Frogmouth sculpin
Icelinus quadriseriatus (Lockington, 1880)	Yellowchin sculpin
Icelinus tenuis Gilbert, 1890	Spotfin sculpin
Jordania zonope Starks, 1895	Longfin sculpin
Leiocottus hirundo Girard, 1856	Lavender sculpin
Leptocottus armatus Girard, 1854	Pacific staghorn sculpin*
Nautichthys oculofasciatus (Girard, 1857)	Sailfin sculpin*
Oligocottus maculosus Girard, 1856	Tidepool sculpin
Oligocottus rimensis (Greeley, 1899)	Saddleback sculpin
Oligocottus rubellio (Greeley, 1899)	Rosy sculpin
Oligocottus snyderi Greeley, *in* Jordan and Evermann, 1898	Fluffy sculpin
Orthonopias triacis Starks and Mann, 1911	Snubnose sculpin
Paricelinus hopliticus Eigenmann and Eigenmann, 1899	Thornback sculpin
Radulinus asprellus Gilbert, 1890	Slim sculpin
Radulinus boleoides Gilbert, *in* Jordan and Evermann, 1898	Darter sculpin
Radulinus vinculus Bolin, 1950	Smoothgum sculpin
Rhamphocottus richardsonii Günther, 1874	Grunt sculpin*
Scorpaenichthys marmoratus Girard, 1854	Cabezon
Synchirus gilli Bean, 1890	Manacled sculpin
Zesticelus profundorum (Gilbert, 1896)	Flabby sculpin

Cynoglossidae (Tonguefishes)

Symphurus atricauda (Jordan and Gilbert, 1880)	California tonguefish*

Diodontidae (Porcupinefishes)

Chilomycterus affinis Günther, 1870	Pacific burrfish*
Diodon hystrix Linnaeus, 1758	Spotted porcupinefish

Embiotocidae (Surfperches)

Amphistichus argenteus Agassiz, 1854	Barred surfperch
Amphistichus koelzi (Hubbs, 1933)	Calico surfperch

129

Amphistichus rhodoterus (Agassiz, 1854) — Redtail surfperch
Brachyistius frenatus Gill, 1862 — Kelp perch
Cymatogaster aggregata Gibbons, 1854 — Shiner perch*
Cymatogaster gracilis Tarp, 1952 — Island perch
Damalichthys vacca Girard, 1854 — Pile perch
Embiotoca jacksoni Agassiz, 1853 — Black perch*
Embiotoca lateralis Agassiz, 1854 — Striped seaperch
Hyperprosopon anale Agassiz, 1861 — Spotfin surfperch
Hyperprosopon argenteum Gibbons, 1854 — Walleye surfperch
Hyperprosopon ellipticum (Gibbons, 1854) — Silver surfperch
Hypsurus caryi (Agassiz, 1853) — Rainbow seaperch
Micrometrus aurora (Jordan and Gilbert, 1880) — Reef perch
Micrometrus minimus (Gibbons, 1854) — Dwarf perch*
Phanerodon atripes (Jordan and Gilbert, 1880) — Sharpnose seaperch
Phanerodon furcatus Girard, 1854 — White seaperch
Rhacochilus toxotes Agassiz, 1854 — Rubberlip seaperch
Zalembius rosaceus (Jordan and Gilbert, 1880) — Pink seaperch

Engraulidae (Anchovies)
Anchoa compressa (Girard, 1858) — Deebody anchovy*
Anchoa delicatissima (Girard, 1854) — Slough anchovy
Anchoviella miarchus (Jordan and Gilbert, 1882) — Slim anchovy
Centengraulis mysticetus (Günther, 1867) — Anchoveta
Engraulis mordax Girard, 1854 — Northern anchovy

Ephippidae (Spadefishes)
Chaetodipterus zonatus (Girard, 1858) — Pacific spadefish*

Gerreidae (Mojarras)
Eucinostomus argenteus Baird and Girard, *in* Baird, 1855 — Silver mojarra
Eucinostomus gracilis (Gill, 1862) — Pacific flagfin mojarra*

Gobiesocidae (Clingfishes)
Gobiesox eugrammus Briggs, 1955 — Lined clingfish
Gobiesox maeandricus (Girard, 1858) — Northern clingfish
Gobiesox papillifer Gilbert, 1890 — Bearded clingfish
Gobiesox rhessodon Smith, 1811 — California clingfish
Rimicola dimorpha Briggs, 1955 — Southern clingfish
Rimicola eigenmanni (Gilbert, 1890) — Slender clingfish*
Rimicola muscarum (Meek and Pierson, 1895) — Kelp clingfish

Gobiidae (Gobies)
Clevelandia ios (Jordan and Gilbert, 1882) — Arrow goby

Coryphopterus nicholsii (Bean, 1882)	Blackeye goby*
Eucyclogobius newberryi (Girard, 1856)	Tidewater goby
Gillichthys mirabilis Cooper, 1864	Longjaw mudsucker*
Gobionellus longicaudus (Jenkins and Evermann, 1889	Longtail goby
Ilypnus gilberti (Eigenmann and Eigenmann, 1889)	Cheekspot goby
Lepidogobius lepidus (Girard, 1858)	Bay goby
Lethops connectens Hubbs, 1926	Halfblind goby
Lythrypnus dalli (Gilbert, 1890)	Bluebanded goby
Lythrypnus zebra (Gilbert, 1890)	Zebra goby
Quietula ycauda (Jenkins and Evermann, 1889)	Shadow goby
Tridentiger trigonocephalus (Gill, 1858)	Chameleon goby
Typhlogobius californiensis Steindachner, 1880	Blind goby*

Hemiramphidae (Halfbeaks)

Euleptorhamphus longirostris (Cuvier, 1829)	Ribbon halfbeak*
Hemiramphus saltator Gilbert and Starks, 1904	Longfin halfbeak
Hyporhamphus rosae (Jordan and Gilbert, 1880)	California halfbeak
Hyporhamphus unifasciatus (Ranzani, 1842)	Common halfbeak

Hexagrammidae (Greenlings)

Hexagrammos decagrammus (Pallas, 1810)	Kelp greenling
Hexagrammos lagocephalus (Pallas, 1810)	Rock greenling
Ophiodon elongatus Girard, 1854	Lingcod
Oxylebius pictus Gill, 1862	Painted greenling*
Pleurogrammus monopterygius (Pallas, 1810)	Forktail greenling

Kyphosidae (Sea chubs)

Girella nigricans (Ayres, 1860)	Opaleye
Hermosilla azurea Jenkins and Evermann, 1889	Zebraperch
Kyphosus analogus (Gill, 1862)	Striped sea chub
Medialuna californiensis (Steindachner, 1875)	Halfmoon*

Labridae (Wrasses)

Halichoeres semicinctus (Ayres, 1859)	Rock wrasse*
Oxyjulis californica (Günther, 1861)	Señorita*
Pimelometopon pulchrum (Ayres, 1854)	California sheephead

Liparididae (Snailfishes)

Careproctus melanurus Gilbert, 1892	Blacktail snailfish
Careproctus osborni (Townsend and Nichols, 1925)	Pink snailfish
Liparis florae (Jordan and Starks, 1895)	Tidepool snailfish

Liparis fucensis Gilbert, *in* Jordan and Starks, 1895	Slipskin snailfish
Liparis mucosus Ayres, 1855	Slimy snailfish
Liparis pulchellus Ayres, 1855	Showy snailfish*
Liparis rutteri (Gilbert and Snyder, *in* Jordan and Evermann, 1898)	Bandtail snailfish
Lipariscus nanus Gilbert, 1915	Pygmy snailfish
Nectoliparis pelagicus Gilbert and Burke, 1912	Pelagic snailfish
Paraliparis albescens Gilbert, 1915	Phantom snailfish
Paraliparis caudatus Gilbert, 1915	Humpback snailfish
Paraliparis cephalus Gilbert, 1891	Swellhead snailfish
Paraliparis dactylosus Gilbert, 1896	Red snailfish
Paraliparis deani Burke, 1912	Pallid snailfish
Paraliparis mento Gilbert, 1892	Bulldog snailfish
Paraliparis rosaceus Gilbert, 1890	Rosy snailfish
Paraliparis ulochir Gilbert, 1896	Broadfin snailfish
Rhinoliparis attenuatus Burke, 1912	Slim snailfish
Rhinoliparis barbulifer Gilbert, 1896	Longnose snailfish

Mugilidae (Mullets)
Mugil cephalus Linnaeus, 1758	Striped mullet*

Mullidae (Goatfishes)
Mulloidichthys dentatus (Gill, 1862)	Mexican goatfish*

Muraenidae (Morays)
Gymnothorax mordax (Ayres, 1859)	California moray*

Ophichthidae (Snake eels)
Myrophis vafer Jordan and Gilbert, 1883	Pacific worm eel
Ophichthus triserialis (Kaup, 1856)	Spotted snake eel*
Ophichthus zophochir Jordan and Gilbert, 1882	Yellow snake eel

Ophidiidae (Cusk-eels)
Otophidium scrippsi Hubbs, 1916	Basketweave cusk-eel*
Otophidium taylori (Girard, 1858)	Spotted cusk-eel

Osmeridae (Smelt)
Allosmerus elongatus (Ayres, 1854)	Whitebait smelt
Hypomesus pretiosus (Girard, 1854)	Surf smelt*
Hypomesus transpacificus McAllister, 1963	Delta smelt
Spirinchus starksi (Fisk, 1913)	Night smelt
Spirinchus thaleichthys (Ayres), 1860	Longfin smelt
Thaleichthys pacificus (Richardson, 1836)	Eulachon

Ostraciontidae (Boxfishes)
  Ostracion diaphanum Bloch and Schneider, 1801   Spiny boxfish*

Pholididae (Gunnels)
  Apodichthys flavidus Girard, 1854   Penpoint gunnel
  Pholis clemensi Rosenblatt, 1964   Longfin gunnel
  Pholis laeta (Cope, 1873)   Crescent gunnel
  Pholis ornata (Girard, 1854)   Saddleback gunnel*
  Pholis schultzi Hubbs, in Schultz, 1931   Red gunnel
  Ulvicola sanctaerosae Gilbert and Starks in
    Gilbert, 1897   Kelp gunnel*
  Xererpes fucorum (Jordan and Gilbert, 1880)   Rockweed gunnel

Pleuronectidae (Righteye flounders)
  Atheresthes stomias (Jordan and Gilbert, 1880) Arrowtooth flounder
  Clidoderma asperrimum (Temminck and Schlegel,
    in Siebold, 1846)   Knobby flounder
  Embassichthys bathybius (Gilbert, 1890)   Deepsea sole
  Eopsetta jordani (Lockington, 1879)   Petrale sole
  Glyptocephalus zachirus Lockington, 1879   Rex sole
  Hippoglossoides elassodon Jordan and Gilbert,
    1880   Flathead sole
  Hippoglossus stenolepis Schmidt, 1904   Pacific halibut
  Hypsopsetta guttulata (Girard, 1856)   Diamond turbot*
  Isopsetta isolepis (Lockington, 1880)   Butter sole
  Lepidopsetta bilineata (Ayres, 1855)   Rock sole
  Lyopsetta exilis (Jordan and Gilbert, 1880)   Slender sole
  Microstomus pacificus (Lockington, 1879)   Dover sole
  Parophrys vetulus Girard, 1854   English sole
  Platichthys stellatus (Pallas, 1787)   Starry flounder
  Pleuronichthys coenosus Girard, 1854   C-O turbot
  Pleuronichthys decurrens Jordan and Gilbert, 1881   Curlfin turbot
  Pleuronichthys ritteri Starks and Morris, 1907   Spotted turbot
  Pleuronichthys verticalis Jordan and Gilbert, 1880 Hornyhead turbot
  Psettichthys melanostictus Girard, 1854   Sand sole
  Reinhardtius hippoglossoides (Walbaum, 1792)   Greenland halibut

Pomacentridae (Damselfishes)
  Chromis punctipinnis (Cooper, 1863)   Blacksmith
  Hypsypops rubicundus (Girard, 1854)   Garibaldi*

Pomadasyidae (Grunts)
  Anisotremus davidsonii (Steindachner, 1875)   Sargo

133

Xenistius californiensis (Steindachner, 1875)   Salema*

Priacanthidae (Bigeyes)
Pseudopriacanthus serrula (Gilbert, 1890)   Catalufa*

Sciaenidae (Croakers)
Cheilotrema saturnum (Girard, 1858)   Black croaker*
Cynoscion nobilis (Ayres, 1860)   White seabass
Cynoscion parvipinnis Ayres, 1862   Shortfin corvina
Genyonemus lineatus (Ayres, 1855)   White croaker
Menticirrhus undulatus (Girard, 1854)   California corbina
Roncador stearnsii (Steindachner, 1875)   Spotfin croaker
Seriphus politus Ayres, 1860   Queenfish
Umbrina roncador Jordan and Gilbert, 1882   Yellowfin croaker

Scorpaenidae (Scorpionfishes)
Scorpaena guttata Girard, 1854   California scorpionfish
Scorpaenodes xyris (Jordan and Gilbert, 1882)   Rainbow scorpionfish
Sebastes aleutianus (Jordan and Evermann, 1898)   Blackthroat rockcod
Sebastes alutus (Gilbert, 1890)   Pacific ocean perch
Sebastes atrovirens (Jordan and Gilbert, 1880)   Kelp rockcod
Sebastes auriculatus Girard, 1854   Brown rockcod
Sebastes aurora (Gilbert, 1890)   Aurora rockcod
Sebastes babcocki (Thompson, 1915)   Redbanded rockcod
Sebastes borealis Barsukov, 1970   Shortraker rockcod
Sebastes brevispinis (Bean, 1884)   Silvergray rockcod
Sebastes carnatus (Jordan and Gilbert, 1880)   Gopher rockcod
Sebastes caurinus Richardson, 1845   Copper rockcod
Sebastes chlorostictus (Jordan and Gilbert, 1880)   Greenspotted rockcod
Sebastes chrysomelas (Jordan and Gilbert, 1881)   Black-and-yellow rockcod
Sebastes constellatus (Jordan and Gilbert, 1880)   Starry rockcod
Sebastes crameri (Jordan, *in* Gilbert, 1897)   Darkblotched rockcod
Sebastes dallii (Eigenmann and Beeson, 1894)   Calico rockcod
Sebastes diploproa (Gilbert, 1890)   Splitnose rockcod
Sebastes elongatus Ayres, 1859   Greenstriped rockcod
Sebastes ensifer Chen, 1971   Swordspine rockcod
Sebastes entomelas (Jordan and Gilbert, 1880)   Widow rockcod
Sebastes eos (Eigenmann and Eigenmann, 1890)   Pink rockcod
Sebastes flavidus (Ayres, 1863)   Yellowtail rockcod
Sebastes gilli (Eigenmann, 1891)   Bronzespotted rockcod
Sebastes goodei (Eigenmann and Eigenmann, 1890)   Chilipepper

Sebastes helvomaculatus Ayres, 1859	Rosethorn rockcod
Sebastes hopkinsi (Cramer, 1895)	Squarespot rockcod
Sebastes jordani (Gilbert, 1896)	Shortbelly rockcod
Sebastes lentiginosus Chen, 1971	Freckled rockcod
Sebastes levis (Eigenmann and Eigenmann, 1889)	Cowcod
Sebastes macdonaldi (Eigenmann and Beeson, 1893)	Mexican rockcod
Sebastes maliger Jordan and Gilbert, 1880)	Quillback rockcod
Sebastes melanops Girard, 1856	Black rockcod
Sebastes melanostomus (Eigenmann and Eigenmann, 1890)	Blackgill rockcod
Sebastes miniatus (Jordan and Gilbert, 1880)	Vermilion rockcod
Sebastes mystinus (Jordan and Gilbert, 1881)	Blue rockcod
Sebastes nebulosus Ayres, 1854	China rockcod*
Sebastes nigrocinctus Ayres, 1859	Tiger rockcod
Sebastes ovalis (Ayres, 1863)	Speckled rockcod
Sebastes paucispinis Ayres, 1854	Bocaccio
Sebastes phillipsi (Fitch, 1964)	Chameleon rockcod
Sebastes pinniger (Gill, 1864)	Canary rockcod
Sebastes proriger (Jordan and Gilbert, 1880)	Redstripe rockcod
Sebastes rastrelliger (Jordan and Gilbert, 1880)	Grass rockcod*
Sebastes reedi (Westrheim and Tsuyuki, 1967)	Yellowmouth rockcod
Sebastes rosaceus Girard, 1854	Rosy rockcod
Sebastes rosenblatti Chen, 1971	Greenblotched rockcod
Sebastes ruberrimus (Cramer, 1895)	Yelloweye rockcod
Sebastes rubrivinctus (Jordan and Gilbert, 1880)	Flag rockcod
Sebastes rufinanus Lea and Fitch, 1972	Dwarf-red rockcod
Sebastes rufus (Eigenmann and Eigenmann, 1890)	Bank rockcod
Sebastes saxicola (Gilbert, 1890)	Stripetail rockcod
Sebastes semicinctus (Gilbert, 1897)	Halfbanded rockcod
Sebastes serranoides (Eigenmann and Eigenmann, 1890)	Olive rockcod
Sebastes serriceps (Jordan and Gilbert, 1880)	Treefish
Sebastes simulator Chen, 1971	Pinkrose rockcod
Sebastes umbrosus (Jordan and Gilbert, 1882)	Honeycomb rockcod
Sebastes vexillaris (Jordan and Gilbert, 1880)	Whitebelly rockcod
Sebastes wilsoni (Gilbert, 1915)	Pygmy rockcod
Sebastes zacentrus (Gilbert, 1890)	Sharpchin rockcod
Sebastolobus alascanus Bean, 1890	Shortspine thornyhead
Sebastolobus altivelis Gilbert, 1896	Longspine thornyhead

Scytalinidae (Graveldivers)

Scytalina cerdale Jordan and Gilbert, 1880	Graveldiver*

Serranidae (Sea basses)

Epinephelus analogus Gill, 1863	Spotted cabrilla
Epinephelus niveatus (Valenciennes, *in* Cuvier and Valenciennes, 1828)	Snowy grouper
	Splittail bass
Hemanthias peruanus (Steindachner, 1875)	Gulf grouper
Mycteroperca jordani (Jenkins and Evermann, 1889)	Broomtail grouper
Mycteroperca xenarcha Jordan, 1888	Kelp bass
Paralabrax maculatofasciatus (Steindachner, 1868)	Spotted sand bass*
Paralabrax nebulifer (Girard, 1854)	Barred sand bass
Roccus saxatilis (Walbaum, 1792)	Striped bass
Stereolepis gigas Ayres, 1859	Giant sea bass

Sparidae (Porgies)

Calamus brachysomus (Lockington, 1880)	Pacific porgy*

Stichaeidae (Pricklebacks)

Anoplarchus insignis Gilbert and Burke, 1912	Slender cockscomb
Anoplarchus purpurescens Gill, 1861	High cockscomb
Askoldia species	Sixspot prickleback
Cebidichthys violaceus (Girard, 1854)	Monkeyface prickleback
Chirolophis nugator (Jordan and Williams, *in* Jordan and Starks, 1895)	Mosshead warbonnet*
Lumpenus sagitta Wilimovsky, 1956	Snake shanny
Phytichthys chirus (Jordan and Gilbert, 1880)	Ribbon prickleback
Plagiogrammus hopkinsii Bean, 1893	Crisscross prickleback*
Plectobranchus evides Gilbert, 1890	Bluebarred prickleback
Poroclinus rothrocki Bean, 1890	Whitebarred prickleback
Stichaeopsis species	Masked prickleback
Xiphister atropurpureus (Kittlitz, 1858)	Black prickleback
Xiphister mucosus (Girard, 1858)	Rock prickleback*

Stromateidae (Butterfishes)

Peprilus simillimus (Ayres, 1860)	Pacific pompano*

Syngnathidae (Pipefishes and seahorses)

Hippocampus ingens Girard, 1858	Pacific seahorse
Syngnathus arctus (Jenkins and Evermann, 1889)	Snubnose pipefish
Syngnathus auliscus (Swain, 1882)	Barred pipefish
Syngnathus californiensis Storer, 1845	Kelp pipefish*
Syngnathus leptorhynchus Girard, 1854	Bay pipefish

Tetraodontidae (Puffers)

Lagocephalus lagocephalus (Linnaeus, 1758)	Oceanic puffer

Sphoeroides annulatus (Jenyns, 1842)   Bullseye puffer*

Sphoeroides lobatus (Steindachner, 1870)   Longnose puffer

Trichodontidae (Sandfishes)

Trichodon trichodon (Tilesius, 1811)   Pacific sandfish*

APPENDIX II

Status of Collections of Tidepool and
Nearshore Fishes of California

The growth of collections of nearshore fishes in California
dates to the founding of the California Academy of Sciences
(CAS) and Stanford University. Much of the ichthyological
material collected prior to 1900 was deposited either at the
Philadelphia Academy of Natural Sciences or the National Museum of Natural History in Washington, D.C.; after 1900, however, collections began to accumulate in California's institutions.

The largest fish collection in California is at the California
Academy of Sciences in San Francisco. This is our oldest collection of fishes, although most of the Academy's early pre-1900
collections were destroyed by the San Francisco earthquake on
April 18, 1906. Today it includes many of the smaller Bay Area
collections as well as the large Stanford and George Vanderbilt
ones. The CAS fish collection is primarily tropical reef forms,
but its holdings of Californian nearshore forms are extensive,
particularly for the northern half of the state.

Since its founding in 1891, Stanford University has pioneered
in research on fishes and amassed tremendous collections. Unfortunately, the university decided to terminate its active role in
the care of systematic collections, and in 1970 all of its fish
collections were transferred to the California Academy of Sciences.

Scripps Institution of Oceanography (SIO) in La Jolla, now a
branch of the University of California, San Diego, began in 1892
as a field center and has grown into one of our most influential
institutions with important marine facilities. Its fish collection is
very large, with major holdings of world-wide deepsea representatives, eastern Pacific tropical and subtropical nearshore fish,
and fishes of California.

In 1910 the University of Southern California established a
marine biological station at Venice, California. Collections of
nearshore fishes were made principally by dredging from the boat
Anton Dohrn between Point Conception and the Mexican

boundary, including the Channel Islands. These collections, which were reported on by Albert B. Ulrey and Paul O. Greeley, were subsequently destroyed when the Venice marine station burned down prior to 1920. Between 1931 and 1941 the University's collection-oriented activities, through the Allan Hancock Foundation, were principally nearshore, but since 1948 emphasis has been concentrated on the deepsea. Today, all of the fishes previously deposited in the Hancock Foundation have been transferred to the Natural History Museum of Los Angeles County for archival storage.

The California Department of Fish and Game, which in 1917 established a marine laboratory in southern California, has grown into a major statewide institution that actively pursues research on fishes and makes recommendations for management of our resources. Although fishes are regularly collected by the Fish and Game staff, they are deposited, after study, at California's institutions which are responsible for archival storage.

Chiefly through the efforts of Carl L. Hubbs, significant collections of Californian nearshore fishes were deposited between 1920 and 1935 at the University of Michigan Museum of Zoology.

Probably the most significant collection of California's nearshore fishes is that of the University of California, Los Angeles (UCLA). This collection, which began in 1948, grew in response to the research activities of Boyd W. Walker and his legion of students. Much of the UCLA fish collection has been transferred to the Natural History Museum of Los Angeles County for archival storage.

The Natural History Museum of Los Angeles County (LACM) has the second-largest collection of fishes in California, with emphasis on future growth defined and restricted to the eastern Pacific Ocean corridor from Alaska to Chile. The collection has grown from a few thousand specimens in 1959 to nearly 2 million today. It has very large holdings of California's nearshore fishes.

Several other fish collections in California have representatives of nearshore species. Basically teaching or small research collections, these include: California Polytechnic College, Pomona;

California Polytechnic College, San Luis Obispo; California State University, Fullerton; California State University, Humboldt; California State University, Long Beach; Occidental College, Los Angeles; Santa Barbara Natural History Museum; University of California, Irvine; University of California, Santa Barbara; University of San Diego.

Through the Association of Systematic Collections, there are plans to develop resource centers for the long-term (archival) storage of fishes (and other types of collections as well) throughout the United States. Certain institutions would be selected as centers for this type of storage, so that the collections would be given maximum protection and care for the greatest period of time. Three institutions in California are under consideration: CAS, LACM, SIO. Some of the smaller fish collections already have been transferred to the proposed archival resource centers. Because of the size and scope of California's three largest fish collections (CAS, LACM, SIO), we suggest that you contact the respective staffs of these institutions for information regarding fishes.

Although state law prohibits the collection of marine invertebrates from the high tide mark to 1000 yards (914 meters) offshore, no prohibition is applied to intertidal fishes. However, a scientific collecting permit is required to take any kind of fish other than by hand or by normal hook-and-line methods, or juveniles of game species. Of all the species that we list in this book, only one, the garibaldi, is fully protected and may not be taken or possessed. We suggest that you contact the Department of Fish and Game before making any collections.

California's Fossil Record of Tidepool and Nearshore Fishes

Fishes from rocky habitat have seldom left a fossil record, a fact often overlooked by paleontologists with little ichthyological background. Because of this, various fossil remains have been assigned erroneously to genera and/or families of fishes which are strictly rock-dwelling or reef-inhabiting forms. Fortunately, this has not occurred in North American paleontological literature as often as in other parts of the world.

141

California has an abundance of well-preserved marine strata which give an almost continuous record of the fish faunas inhabiting the coastal waters during the most recent 60 million years of the earth's history. Identifiable remains come in a wide variety of forms, ranging from intact three-dimensional fishes to isolated items such as bones, otoliths, scales, and teeth.

Intact three dimensional fossils are the most easily identified, because recognition characters usually are present and in place. The best of these remains are found in clays, but siltstones and diatomites sometimes contain skulls or entire skeletons in three-dimensional form. Typically, however, fish remains found in bedded diatomites, shales, and sandstones are compressed two-dimensional forms, sometimes called "skeletal imprints" or simply "imprints." Perhaps 100 kinds of two- and three-dimensional fishes have been found in fossil deposits in California, but only about one-fourth of these may represent nearshore forms.

Sandy and silty strata are most likely to contain isolated remains, but a surprising number of these can be identified if they are not too badly worn or fragmented. Certain vertebrae from cusk-eels, pipefishes, cods, and several other species are unique and distinctive, as are opercular spines of numerous sculpins, body spines of porcupinefishes and their relatives, fin spines of triggerfishes, and skull bones of some triglids. Teeth from wrasses, sparids, triggerfishes, atherinids, morays, and embiotocid perch have all been identified from Californian fossil deposits, as have scales from herrings, pipefishes, jacks, poachers, and cods. Some of the most abundant of identifiable fish remains, however, are otoliths. Between 350 and 400 kinds of otoliths have been found in marine and freshwater deposits within the boundaries of the state, some as old as the Cretaceous period. Fishes belonging to 28 of the 54 families covered in this book have left their otoliths in one or more of these deposits.

Unfortunately, with most fragmentary remains there is no way to determine if one type is from the same species as some other type found in the same deposit. Thus an isolated otolith, scale, tooth, or vertebra may actually be from the same extinct species, but without an intact specimen for comparison there is no way to determine this, and each of the four elements could be ascribed to a different species. Thus a fossil fauna could appear to be more affluent than was actually the case.

Diatomites, shales, and sandstones which contain fish remains occur primarily in southern California and are mostly Miocene in age. The best localities are near San Clemente, San Pedro, Walteria, Santa Monica, Pomona, Burbank, Thousand Oaks, Moorpark, Gaviota, Lompoc, Surf, and Taft. Fossiliferous sandy and silty beds are found throughout the entire state, but are especially abundant along the coast. San Diego County is best for workable Eocene strata, while Kern and Orange counties contain the richest fish-bearing Miocene outcrops. Pliocene and Pleistocene exposures are ubiquitous, but the most productive beds can be found in Humboldt, Monterey, San Luis Obispo, Kings, Santa Barbara, Ventura, Los Angeles, Orange, San Diego, Riverside, and Imperial counties. Friable Oligocene deposits which contain recognizable fish remains are extremely scarce in California, especially in surface outcrops.

A Brief History of Ichthyology as It Concerns
California's Tidepool and Nearshore Fishes

Ichthyology on the Pacific Coast of North America dates to two explorations by the Russians and British. Petrus S. Pallas reported on the collections of fishes initiated by the Russian government in 1810-1811, and Albert Günther reported on those fishes taken by the H.M.S. *Plumper* in 1857 off Canada. Prior to 1850 only 25 of the 349 species in our checklist had been described, but not from off California. In fact, the tidepools of California had not been explored, and their fish fauna was still unknown.

Ichthyology in the coastal waters off California dates to the arrival of a young Massachusetts physician, Dr. William O. Ayres (1817-1891), who was among the forty-niners lured to California by the search for gold. Ayres is considered the first of California's ichthyologists, and was one of the founders of the California Academy of Sciences, in San Francisco, where the early collections were deposited. (They were destroyed in the earthquake of April 18, 1906.)

The first major contributions on the kinds of tidepool and nearshore fishes were made by William Ayres and Charles F. Girard (1822-1895). Of the 116 species of fishes described and added to the list of tidepool and nearshore fauna between 1850 and 1875, 51 were proposed by Girard and 28 by Ayres, and they are still recognized today. Much of the material named by Girard was collected during three major systematic surveys of the flora and fauna of the western half of the United States and the Pacific Coast of North America: the "Pacific Railroad Survey", "Exploration of the Mexican Boundary," and "Exploration of the Western Half of the United States." Ayres gathered fishes from along the central California coast and from the San Francisco markets. Louis Agassiz (1807-1873) pioneered in the discovery of the surfperches and their remarkable feat of giving birth to living young. Others who observed the viviparous nature of the surfperches include William P. Gibbons, A. C. Jackson, J. K. Lord, and Thomas H. Webb. Gibbons (1812-1897), Friedrich H. von Kittlitz (1799-1874), Theodore N. Gill (1837-1914), James G. Cooper (1830-1902), and Franz Steindachner (1834-1919) described fishes that are still recognized today.

During the next twenty-five years, from 1875 to 1900, 149 additional nearshore fishes were described; 101 of these were named either by David S. Jordan (1851-1931) or Charles H. Gilbert (1859-1928), or in collaboration with their ichthyological associates at Stanford Univeristy.The contribution to the development of ichthyology by these two distinguished biologists, and their colleagues at Stanford, was so great that their influence continues today through second- and third-generation scientists. Their active and dominant role in the studies on Californian fishes and the ichthyology of North America resulted in the publication in 1896-1900 of *Fishes of North and Middle America* by David Jordan and Barton W. Evermann (1853-1932), which today remains a standard text for icthyologists.

Ten other prominent scholars were active during this period: Tarleton H. Bean (1846-1916), Frank Cramer (1861-1948), Carl H. Eigenmann (1863-1927) and his wife Rosa Smith (1859-1947), Arthur W. Greeley (1875-1904), Oliver P. Jenkins (1850-1934), William N. Lockington (1842?-1902), John O. Snyder (1867-

1943), Edwin C. Starks (1867-1932), and Joseph Swain (1857-1927).

By 1900, 280 of the 349 tidepool and nearshore fishes in our list had been discovered and named, establishing a reasonable knowledge of the fish fauna of California. Although new species continued to be proposed, investigations turned from descriptive ichthyology to contributions on anatomy, checklists, general treatises, phylogenetic revisionary works, and attempts to treat part or all of California's fishes. Only 26 kinds of nearshore fish species that are recognized today were proposed between 1900 and 1925. Many of these 26 were described in revisionary contributions by Charles H. Gilbert. More important, the initial attempts to report on the marine fishes of southern California began. *The Marine Fishes of Southern California* by Edwin Starks and Earl L. Morris was the first to define the fish fauna of this geographic area of the state, but the manuscript was destroyed in the 1906 San Francisco earthquake. It was issued subsequently as a checklist (1907). Albert B. Ulrey and Paul O. Greeley made several small contributions to the list of southern California marine fish fauna (1923, 1924, 1929). The third contribution on the fishes of this area was a rather uninspired treatment by Percy Barnhart (1936).

The next twenty-five years produced only 9 species descriptions, but in 1948 the first edition of the *Checklist of the Fishes of California* was mimeographed by Carl L. Hubbs and W. I. Follett. This has been a standard reference to the fish fauna of the state, although in its fourteenth edition it still remains in mimeographed form. The influence and contributions of Hubbs and Follett place them in the forefront of California's living ichthyologists. The classic study by Rolf L. Bolin (1901-1973) on the sculpins, family Cottidae, was published in 1944, and the early works of Earl S. Herald (1914-1973) were published in the early 1940s.

Since 1950, 24 species have been proposed, usually in significant revisionary papers. Some of these are the studies on the kelpfishes, family Clinidae, by Clark Hubbs; the surfperches, family Embiotocidae, by Fred Tarp; the clingfishes, family Gobiesocidae, by John Briggs; and the rockcod, family Scorpaenidae, by Julius Phillips. A number of publications on California's nearshore fish fauna have come since 1950 from the staff of the Department of Fish and Game, including John E. Fitch, Robert N. Lea, Daniel Miller, and Phil Roedel.

Even today several new species of nearshore fishes, recently discovered by scuba methods in the subtidal region, remain to be described. Most of these include small fish, such as gunnels and pricklebacks, that hide in burrows, kelp, or other rocky-type habitats. Fishes living in California's nearshore habitats are important because of their prominence here, their potential as indicators of pollution, and their inclusion in environmental reports. Significant contributions to our knowledge of tidepool and nearshore fishes can be made in life-history and behavioral studies.

APPENDIX III
Helpful References

Bailey, Reeve M., John E. Fitch, Earl S. Herald, Ernest A Lachner, C. C. Lindsey, C. Richard Robins, and W. B. Scott. 1970. A list of common and scientific names of fishes from the United States and Canada. American Fish. Soc., Spec. Publ. 6:1-150.

Baxter, John L. 1966. Inshore fishes of California. California Dept. Fish and Game, Sacramento. 80 pp.

Bolin, Rolf L. 1944. A review of the marine cottid fishes of California. Stanford Ichthyol. Bull. 3(1):1-135.

Briggs, John C. 1955. A monograph of the clingfishes (Order Xenopterygii). Stanford Ichthyol. Bull. 6:1-224.

Burke, Victor. 1930. Revision of the fishes of the family Liparidae. Bull. U.S. Natl. Mus. 150:1-204.

Cannon, Raymond. 1964. How to fish the Pacific Coast. Lane Pub. Co., Menlo Park. 337 pp.

Hart, J. L. 1973. Pacific fishes of Canada. Fish. Res. Bd. of Canada, Ottawa. 740 pp.

Herald, Earl S. 1961. Living fishes of the world. Doubleday and Co., New York. 304 pp.

_____. 1972. Fishes of North America. Doubleday and Co., New York. 254 pp.

Hubbs, Clark. 1952. A contribution to the classification of the blennioid fishes of the family Clinidae, with a partial revision of the eastern Pacific forms. Stanford Ichthyol. Bull. 4(2):41-165.

Marshall, Norman B. 1965. The life of fishes. World Publishing Co., New York. 402 pp.

McAllister, Don E. 1963. A revision of the smelt family, Osmeridae. Bull. Natl. Mus. of Canada 191:1-53.

Miller, Daniel J., and Robert N. Lea. 1972. Guide to the coastal marine fishes of California. California Dept. Fish and Game, Fish Bull. 157:1-235.

Norman, J. R., and P. H. Greenwood. 1963. A history of fishes. Ernest Benn Ltd., London. 398 pp.

Phillips, Julius B. 1957. A review of the rockfishes of California
(Family Scorpaenidae). California Dept. Fish and Game, Fish
Bull. 104:1-158.

Quast, Jay C., and Elizabeth L. Hall. 1972. List of fishes of
Alaska and adjacent waters with a guide to some of their
literature. Natl. Mar. Fish. Serv., Tech. Rept. SSRF-658:1-47.

Turner, Charles H., and Jeremy C. Sexsmith. 1964. Marine baits of
California. California Dept. Fish and Game, Sacramento.
71 pp.

Zim, Herbert S., and H. H. Shoemaker. 1956. Fishes: A guide to
familiar American species. Simon and Schuster, New York.
160 pp.

benthos: animals and plants which live on the bottom of the sea.

calcareous: containing or like calcium carbonate; seashells and fish otoliths are typical examples of calcareous objects.

cirrus: a small fleshy appendage or filament, usually found in the head region.

coelenterates: group name for some gelatinous-bodied animals such as jellyfish, sea anemones, sea pens, and similar creatures.

crustaceans: group name for joint-legged animals with horny or chitinous exoskeletons; typical crustaceans are crabs, lobsters, shrimp, amphipods, copepods, mysids, euphausiids, and isopods.

ctenoid: a type of fish scale, having small stiff projections on one edge, usually rough to the touch.

diatomite: a fine siliceous earth composed mainly of cell walls of one-celled algae (diatoms).

encircling net: any net which is used to surround fish schools; typical encircling nets are purse seines, lamparas, and bait nets.

euryhaline: here intended for fishes which tolerate a wide range of salinities; that is, they are capable of living in salt, brackish, or freshwater.

family: a group or category of plants or animals in which all members have a set of characteristics common to each of them.

fecundity: the egg-bearing capacity of female fishes, differing greatly depending upon the age (size) of the individual.

fish-of-the-year: fish hatched during current year.

genus: the first part of the scientific name of any animal or plant; the first letter of a generic name is always capitalized.

gill cover: the operculum, a flap which protects the gills.

gill net: any net that hangs wall-like in the water and snares fishes which are too large to swim through it. A trammel net is a modified gill net that is most efficient for catching flatfishes.

habitat: the area where a fish lives.

hermaphrodite: an individual with both male and female organs, a normal arrangement in some fishes.

holotype: the unique individual for which a species is named; holotypes typically are retained by museums and other scientific institutions, and often are stored separately from general collections.

ichthyocide: a chemical substance which when introduced into the water will stun or kill fishes; it is illegal to use these chemicals without a permit, and such permits are limited to organizations or individuals involved in research of fish management activities.

ichthyology: the study of fishes; from the greek ichthyos (meaning fish) and ology (the study of).

Indian midden: a refuse heap left by prehistoric indians, usually marking campsites.

intertidal zone: the nearshore area of the beach marked by the action of the tides; literally between the tides.

larvae: the young stages of fishes, usually before they obtain their adult pigment patterns.

lateral line: typically, a row of pored scales which traverses the midside of a fish from the gill opening to the base of the tail; it may be interrupted, curved, branched, or absent, or there may be more than one, depending upon species.

leptocephalus: a transparent, elongate, or leaf-shaped larval stage of bonefishes, eels, and some related families.

Miocene: a period in geologic history which is generally thought to have started about 25 million years ago and ended perhaps 10 million years ago, when the Pliocene began.

mollusks: group name for related shelled and nonshelled animals including snails, clams, squids, octopi, chitons, and others.

nudibranch: a sea slug (mollusk) which lacks a shell and has external respiratory appendages.

omnivorous: eating both plant and animal food indiscriminately.

operculum: the gill cover, the hindmost and uppermost bone of the gill cover.

otoliths: calcareous concretions in the inner ear of a fish, functioning as organs of hearing and balance. There are three pairs of otoliths in the skull of each fish, and these are termed sagittae, lapilli, and asterisci. Otoliths are used by fishery biologists for numerous studies.

pelagic: living in open waters in contrast to bottom or inshore.

photophores: complicated organs for emitting light (luminescing); typically located on sides, heads, or bellies of deep-sea animals, but may be elsewhere on or within the body. Photophores occur in fishes, squids, shrimps, and a wide assortment of other creatures.

plankton: very small, but not necessarily microscopic, aquatic plants and animals usually drifting with currents somewhere above the bottom.

Pleistocene: a period in geologic history which followed the Pliocene period (see below) and ended after the Ice Ages, possibly 9 to 10 thousand years ago.

Pliocene: a period in geologic history which began possibly 10 million years ago, at the end of the Miocene period, and ended about 4 million years ago, when the Pleistocene began.

preopercle: a small triangular bone in front of the operculum or gill cover in fishes.

red tide: a nearshore oceanic phenomenon wherein small one-celled animals reproduce at rapid rates thereby using-up all the available oxygen in the water. Consequently, the animals with gills that require a good supply of oxygen die.

school: a large group of fishes that swim together in a uniform manner.

slurp gun: a gun-shaped device which when triggered creates a suction that effectively collects small living fishes at a distance of several inches from the muzzle.

spawn: the term for reproduction in fishes.

species: the second part of the scientific name of any animal or plant. Both singular and plural, it is often used as a vernacular for type or kind; the first letter of a species name is never capitalized.

substratum: the underlying bottom or floor of the ocean.

viviparous: bringing forth living young, rather than being an egg-layer.

INDEX OF SCIENTIFIC NAMES

INDEX OF COMMON NAMES